WATER

WATER

Towards a Culture
of Responsibility

Antoine Frérot

University of New Hampshire Press

Durham, New Hampshire

University of New Hampshire Press
An imprint of University Press of New England
www.upne.com

First English-language edition © 2011 University of New Hampshire
All rights reserved
Originally published in French as Antoine Frérot, L'eau, pour une culture de la
responsabilité © Éditions Autrement, Paris, 2009

Manufactured in the United States of America

University Press of New England is a member of the Green Press Initiative. The paper
used in this book meets their minimum requirement for recycled paper.

For permission to reproduce any of the material in this book, contact Permissions,
University Press of New England, One Court Street, Suite 250, Lebanon NH 03766;
or visit www.upne.com

Library of Congress Cataloging-in-Publication Data
Frérot, Antoine.
  [L'eau, pour une culture de la responsabilité. English]
  Water : towards a culture of responsibility / Antoine Frérot. — 1st ed.
    p. cm.
  Originally published: Paris : Autrement, 2009.
  ISBN 978-1-58465-987-7 (pbk. : alk. paper) — ISBN 978-1-58465-990-7
  (e-book)
  1. Water resources development. 2. Water-supply—Government policy.
  3. Climatic changes—Economic aspects. I. Title.
  HD1691.F74 2011
  333.91—dc22   2010045819

5 4 3 2 1

"Everyone is responsible to everyone for everything."

DOSTOYEVSKY

CONTENTS

PREFACE

ÁNGEL GURRÍA

OECD Secretary General
Member of the UN Secretary General's Advisory Board
on Water and Sanitation

Financing water for all is the unambiguous title of the report published in preparation for the Third Water Forum held in Japan in 2003: it is not entitled *Financing water for half*. As a former Minister of Economy and Finance for Mexico, I had been invited by my friend Michel Camdessus[1] to take part in a working group on the financing of water infrastructures. We were asked how to finance the commitment undertaken by the international community to decrease by half the number of people without access to drinking water and sanitation by 2015. We were bold enough to state that financing water and sanitation for all of those without it, and not just for half as demanded by the Millennium Goals, was both an accessible target and an exciting task.

The water professionals and finance experts in our group were all profoundly moved by the human distress behind the figures: nearly a billion people have no access to drinking water and more than two billion have no sanitation system. Disease, infant mortality, the daily struggle to obtain water (to the detriment of women's economic activity and children's schooling), these are the human realities for those who lack normal access to the water essential for life and development.

Water is fickle: it will flow freely and unpredictably where it is least needed and is absent where it is vital; it is the major factor in a planetary tragedy. But human fecklessness is also to blame, including wastage, urban, industrial and agricultural pollution, absence of purifying systems and frequently poor democratic and financial management of the sector. Human irresponsibility has helped to make water one of the main human problems for this century. Resolving it has become an international priority, and I have made it a central working concern for OECD economists.

For my part, I am convinced that we must find a way out of the conventional debates so that we do not simply hit a brick wall. Discussions about water are corrupted by indifference or demagogy. A world that prefers providing consumer goods to providing drinking water is a dangerous world for the most needy. The passion surrounding debates about water only reflects the inability of many to put into practice the

----

1. French economist and former president of the International Monetary Fund (IMF) from 1987-2000.

commitments to which they have officially subscribed. The problems we face will only be solved if all those involved, in both public and private sectors, cooperate constructively to achieve their goals.

Antoine Frérot is one of those rare people involved with water who combine a vision of all these challenges and possible futures, managing water and sanitation services on five continents, and has a deep knowledge of all the people involved. As a responsible professional, he refuses to lapse into demagogy or get sidetracked into dead ends. His proposals are informed by industrial and scientific realities. He makes short shrift of false quarrels and identifies solutions which will make it possible for water resources to be in future what they have always been in history: a common concern for all human beings, and a bond between nature and humanity.

Water raises an ethical and a political question. We are only a few years away from 2015. Today it is still possible for us to keep our commitments and to go even faster and further, as we proposed in our group of experts. Water for all in 2025: an accessible dream, a human necessity. In this book Antoine Frérot invites us to mobilise collectively around the known solutions, which we must then put into practice.

INTRODUCTION

A TIME TO TAKE RESPONSIBILITY

Water's symbolic meaning is vast. Every civilisation, every culture, every tradition talks about water and weaves an intricate web of meanings and symbols round it. However, in this maze of multiple echoes and resonances, some key themes emerge: first of all the mystery of our origins. Don't the first words of Genesis state succinctly: "The spirit of God moved over the face of the waters", even before the creation of heaven and earth? Doesn't the Koran also say that before conceiving his creation, God was in a cloud in the middle of the air: "Then he created his throne above the water"? And in Hinduism isn't water a sacred "being", an object of pilgrimage and reverence to the image of the Ganges? At the heart of these stories, beyond the symbolic web of meanings that the history of civilisations has assiduously woven, water symbolises the perfection and purity of a nature before creation, the ineffable trace of a world before the world. Despite the diversity of cultures, water retains that pristine shine, which gives it sacredness. With ritual ablutions in Islam, Christian baptism, Jewish Mikvah, traditional Hindu bathing, water remains the principal means of purification in most religions.

Sovereign and autonomous, water's value lives on beyond the fluctuations and contingencies of human history. The sequence by which it comes to us is the long cycle from earth to heaven that succeeds in creating "the link between spirituality and materiality" (Bergson). It is only with the advent of human wrongdoing that water becomes loaded with negative and potentially destructive connotations. In the Judaeo-Christian tradition, the first of these events is the Flood.

In order to understand the passions aroused by water, we need to grasp its symbolic values, as described by Gaston Bachelard[1], who was both an engineer and a philosopher. Everything begins, he says, with water as a mirror. For human beings looking at themselves and seeking themselves, water becomes a visual reflection. Water is also the symbol of life, because it is life. Deep within every human being, water is associated with images of well-being and nurturing going back to the mother-child relationship. Before birth, the child is bathed in water, the amniotic fluid. That is why, Bachelard continues, "water is milk", echoing Saint-

11

----

1. Gaston Bachelard, *L'eau et les rêves,* 1942.

John Perse's poetic expression, "the calm waters of milk". Water is benef-
icent because it is maternal and protective. Mythical water is the water
of life, but it is also closely connected with death, sometimes conceived
as an infinite voyage over infinite waters. So it is not surprising that life
and death problems are omnipresent in debates about water: pure
water gives life, water's absence or pollution bring death. Water is also
purity. When that purity is destroyed, there is fury. The polluters are pro-
faners. Pure water represents the Good, impure water represents Evil.
In many primitive cultures saving the water is much more than an everyday task: it means saving humanity. In the three monotheistic religions,
water is an essential symbol: Moses, Jesus and Mohammed all perform
water miracles. Beyond cultural specifics, the common feature relating to water is absolute dependence: water is perfect and does not need
humanity, whereas humanity is imperfect and has vital need of water.
This one-way dependence has always governed humanity's relationship with water.

What is new in this century is the challenge to this relationship: at
the risk of grandiloquence, we may say that today water needs us. What
has happened? Firstly, the impact of human activity has profoundly
altered: it has long been insignificant in relation to nature, but today it
has qualitatively changed. In the past humanity was threatened by a
nature that was thought to be invulnerable, but today humanity is a
threat to nature. Certain scientists even speak of a new era, the "anthro-
pocene era". The neologism coined by the Dutchman Paul Crutzen, a
Nobel prizewinner in chemistry (1995), describes the growing impact
of humanity on the biosphere. According to him, this era began around
1800 with the advent of industrial society, characterised by the massive
use of hydrocarbons. By becoming a "planetary geophysical force" we
have become responsible for nature, which is falling more and more into
our own hands. The idea of the "garden", so dear to the humanists of the
past, who saw it as humanity's ability to reproduce and model the forces
of nature in a harmonious way, has now become a necessity: the planet
itself has become a garden and we are its gardeners and guardians.

Here are a few examples, not simply for the sake of frightening ourselves, but to give some measure to the urgency. The availability of fresh
water per world inhabitant has fallen dramatically. It has dropped from
17,000 m^3 per year in 1950 to 7,500 m^3 per year in 1995 and is predicted
to drop to 5,100 m^3 in 2025. This development is mainly due to demographic growth – from now on the planet's population is increasing by
a billion every twelve years, whereas world population only reached a
billion around 1800. And if the population has tripled in one century,
water consumption has multiplied by six. At the same time half the
world's rivers and lakes have become polluted by waste water. Today

Lake Chad only covers 10 % of the area it covered 40 years ago. Why? Partly this is due to a decrease in rainfall, but it is mainly due to human over-exploitation. Demand for irrigation water has quadrupled, rapidly using up a declining resource. The Aral Sea has practically dried up as a result of the diversion of the two rivers that fed it: the Syr-Daria and the Amou-Daria. What remains of it is a vast expanse of dead, over-salty water. We could go on quoting alarming examples, but the conclusion would be the same: the state of water resources is up to us and "we are going to have to learn to converse with a new demanding and insidious partner: scarcity"[2]. This responsibility for our environment and water resources is long-term, lasting well beyond our own lifetime, and concerns future generations. It requires us to act with great care and restraint in order to preserve the long-term conditions for life.

Nevertheless, – and this is certainly one of the most important water issues – responsibility for the present is just as great as the responsibility we have for the future. We need to recall the figures, shocking as they are however often repeated. Nearly a billion people do not have access to drinking water and 2.6 billion have no access to basic sanitation. Every year, 2.2 million people die from diseases linked to the lack of water or poor water quality. We need to act as swiftly as possible. The international community undertook commitments at the Millennium Summit in New York in September 2000 and reaffirmed them at the Johannesburg Earth Summit in 2002: between now and 2015, they said, we must reduce by half the proportion of people without access to water or basic sanitation.

In its 2006 Global Human Development Report, the United Nations Development Programme (UNDP) has drawn up an intermediary balance sheet for what it has been agreed to call the Millennium Goals. The good news is that the world is on the way to achieving the drinking water target – particularly because of the progress made by the two demographic giants, China and India. The bad news is that there are big shortfalls within this global assessment. If we do not speed up the rhythm for the provision of drinking water, 55 countries will fall short of the Millennium Goals, and in 2015, 235 million people will be below the target initially set. As for sanitation, it is lagging sharply behind drinking water in the majority of developing countries: 74 of these countries are failing and 700 million people will suffer from the failure to reach this goal. Sub-Saharan Africa is lagging furthest behind in the provision of drinking water: it will reach its goal a generation later than the due date, and only reach its sanitation target in 2076, that is, two generations

---

2. Economists' Circle and Erik Orsenna, *Un monde de ressources rares*, Paris, Éditions Perrin, 2007, 216 pp.

after the due date. These delays are even more worrying because millions of people are left out of the statistics, so the revealed data do not represent the full extent of the failure. On the other hand, the Millennium Goals should be thought of as a minimum to surpass, not as a final aim. In fact, they are only half a goal because they only aim at halving the population without drinking water and sanitation. Even if these goals were achieved, the world deficit would still be enormous: a total of 800 million human beings would remain without minimal access to drinking water and 1.8 billion without basic sanitation.

What must we do? This dramatic situation is not inevitable. There are solutions but there are also obstacles. First of all, there is a certain silence, a kind of indifference in public opinion. To see this, we need only compare the importance given – albeit perfectly legitimately – to climate change with that given to water access. How many newspaper and TV news headlines are there about climate change? How many "national and international priorities"? The fight against the greenhouse effect – important as it is – has mobilised the general public and Al Gore's DVD was one of the most popular Christmas presents of 2007. Do you know of any film about access to water? This silence is even odder because the consequences of climate change are largely uncertain. On the other hand, lack of water kills people every day. And how is it that aid for development is five times greater for telecommunications than for water? Is it because of the distance separating us from populations suffering from lack of water? Is it compassion fatigue, exhausting our capacity for indignation, with one good cause blotting out another? Is it because we in developed countries are accustomed to enjoying plenty of excellent quality drinking water, so that we cease to realise how precious and vital such a provision is? Perhaps all of these have something to do with it. Perhaps it is also a sign of a certain defeatism, after so many years of hollow declarations, which have had no effect.

Nevertheless, the professional water community (international institutions, governments, NGOs, public or private operators...) have spawned numerous speeches, conferences and debates. Unfortunately, the silence of the media is too often matched by the logorrhoea of the specialists. Already in 1977, at the Mar del Plata Conference, the international community invoked the principle that "everyone has the right to the water necessary for their vital needs". The 1980s were proclaimed the International Drinking Water and Sanitation Decade (1981-1991), with the stated goal being to provide access for the whole planet. Why did these speeches and conferences fail, thus damaging the credibility of commitment to water provision in terms of public opinion? Certainly one of the obstacles was this: water became the object of ideological struggles more often than of concrete reflection about the practical actions to take.

Over the last ten years, controversies have multiplied over free access to water and price increases, the development of water resources and the way to satisfy growing needs, the efforts made to protect these resources and their fair disposal between the various players in the water business, on the role of public development aid for water, on the role of international financial institutions, on dams (which cause large population displacements, loss of agricultural land and reduction in groundwater), on the appropriateness and legitimacy of the private sector involvement... Some have declared that the private sector, always hungrier for profits, have lit upon the most precious of resources – blue gold, as they call it – to engage in a trade some denounce as immoral. Such private operators are dogmatically condemned – with no appeal – by these denouncers, but it is not clear that these "water warriors", as they have sometimes called themselves, serve the cause of water all that well. Indeed, over-simplification is a bad counsellor, for both sides: those who have seen or still see the solution to the water challenge as lying solely in the law of the market are also mistaken. Such quarrels do not do justice either to the complexity or to the urgency of the problem, especially as we know that the private sector is involved in less than 10 % of drinking water provision in the world. "In the great battle to come, water has fewer true allies than it has false friends," says the economist Michel Camdessus.

Of course, water management is a political question *par excellence*. It is therefore necessary to look beyond the effectiveness of different economic models, to their social legitimacy. Here the debates are highly pertinent. But their pertinence ends where over-simplification begins. For water is too complex a question to believe all those for whom it would suffice to pull two or three levers to solve all the problems. The water question is much too difficult to content ourselves with a few simple mantras: effective action requires that we reject Manicheism.

However, these debates are fruitful in that they emphasise that water is a good mirror in which to look at different social responses and understand how a society works. In this respect we can easily show how water reflects the social developments of recent years. As Jacques Lacarrière stresses: "Water, our first mirror, you also reflect our deeds[3]". Firstly, water is the reflection of globalisation; its management illustrates the tension between different intervention levels – local, national and international. Water, which is a collective local service par *excellence*, is also subject to a rise in power of international players: institutions,

---

3. Jacques Lacarrière, "Onde pure, chant premier de la terre", *GÉO*, no.112, June 1988.

financial backers, businesses, but also NGOs, who all help to make it a more and more globalised problem. Water invites a fresh look at models of subsidiarity and decentralisation. Hence water reflects the growing interaction between environmental, economic and social questions – governed by the notion of sustainable development. It is the most complex of sectors: good water management means reconciling the need to protect the resource, economic sustainability and social acceptability. For better or worse, discussions of water also reflect a system of media where the symbolic and emotional often carry more weight than rational argument. Sometimes solutions to problems over water pay the price of its supreme symbolic status. Although the debate in the media is an indispensable lever for rousing public opinion about the problems, it sometimes offers too much scope for over-simplification. Lastly, we can say that water management reflects a new form of governance, involving all the players in the social and political body: states and local communities, private businesses, but also professional associations and NGOs, who are increasingly concerned with water management. Through that management all these players, each with their own logic and mandate, learn day by day, with difficulties it would be pointless to deny, how to coordinate their actions better. Winning the water wager means demonstrating the possibility of effective "governance", seeking the general good by means of decision processes, which will increasingly involve public bodies, businesses and civil society in general.

In fact, water management is one of the first testing grounds for citizen participation – the desire to be involved that we find today in many areas. This emergence of the citizen-consumer amplifies concerns about the provision of water, but also the demand for transparency and a desire not to be simply administered, but to be involved and responsible users. Thus water becomes an important laboratory for the questions of our time, and it would be pretentious and counter-productive to try to offer simplistic answers to the broad questions it raises. On the contrary, the study of water commits us to a new humanism: a new humanism with a broad scope, capable of embracing all that we know, a new humanism capable of putting that knowledge to work in taking decisions and action.

Such is this book's ambition: to distinguish the real problems from the vain polemics, and call for an alliance between responsibility and the will to act. We can and must face all the challenges water creates. Solutions exist; others can be invented or repeated. Putting them into practice is within our reach.

PART 1

THE WATER CENTURY

"To call things by the wrong name", as Albert Camus used to say, "is to add to the misery of the world." And water has suffered too much from the silence or, conversely, the approximations, of buzzwords and overblown language. Discussion on the subject of water sails between two reefs: on the one hand, indifference, and on the other, "doom watching". Both of these in their own way discourage decision and action: indifference, because awareness and a general mobilisation are necessary; doom watching because it leads people to believe that crisis is inevitable, which it certainly is not. The subject of water merits a better treatment. At the very least we should set out to analyse its characteristics and understand the real debates, beyond all the media-grabbing arm waving. Today we have to have a precise analysis that defines the problems so that we can deal with them better. Three challenges stand out: satisfying humanity's growing need for water; stopping the degradation in the quality of our water resources; and bringing the two essential services of drinking water and sanitation to all those who lack them at present.

I. SATISFYING GROWING NEEDS

Water is abundant in nature and human beings have lived for thousands of years without paying much attention to it, apart from those who lived in arid zones or in areas devastated by floods. But today, our water resources appear to be dwindling. While we used to think they were abundant, they are now becoming scarcer even in countries with a relatively heavy rainfall. In Europe, water-use restrictions are regularly imposed in many regions, and water levels are sinking, marking the return of water-resource problems that were thought to have been solved. While these hydrological changes may appear locally, in reality there is no less water on Earth, just more people and more extraction of limited resources. Water is a renewable resource. Unlike oil, which is running out at the rate at which nations burn it, the volume of water on Earth has remained constant – at around 1,400 million km^3* – for 3.4 billion years. It has not varied in all this time, whether during the worst episodes of warming or the coldest periods of glaciation. We have no reason to believe that it will not continue in this way in the future.

The blue planet is aptly named, for water covers 72 % of its surface. Given the volume of water contained in the seas and oceans, 97.5 % of the water present on Earth is salt water, and the remaining 2.5 % is fresh water. But can this 2.5 % be used by humanity? No, because 70 % of fresh water is frozen in the poles and in mountain glaciers, and the remaining 30 % is almost all ground water, stored in deep underground water courses. "If for a moment one could stop all the rivers from flowing, the quantity of water which they contain would scarcely amount to 0.005 % of the total reserve of fresh water[1]," explains Jean-Marie Fritsch, a hydrogeologist at the Development Research Institute (DRI) in South Africa. In the final analysis, the amount of liquid fresh water theoretically available to satisfy all humanity's needs is only 0.3 % of the water on our planet.

However tiny it may be, this quantity is largely sufficient, from a global point of view, but on the local level there may be woeful deficiencies. Fresh water is very unequally distributed in space and time,

----

1. Jean-Marie Fritsch, "La crise de l'eau n'aura pas lieu", *La Recherche*, n° 421, July-Aug. 2008.
* 1 km^3 equals 1 billion cubic meters.

and there is a heavy demand in the arid regions. There is also an imbalance between populated cities and empty countryside, between areas of high consumption and sources of supply. There is unfortunately no overall correlation between population density and availability of water resources. People are crowded into coastal areas and build cities in the heart of arid zones. Even though it is situated in the desert, Dubai, for instance, has seen its demand for water increase by 9 % per year! Las Vegas was created from nothing in one of the driest regions of the United States. The steppes and arid zones receive only 2.2 % of the total flow of water, while 21.5 % of the world's population live there. Conversely, tropical regions, which account for only a sixth of world population, receive half of the global water flow[2].

Unequally distributed geographically, fresh water can be even more unequally distributed over time. The periods when most water is drawn for irrigation are often those when both surface water and underground water resources are at their lowest. There are therefore frequent imbalances between the need for water and its availability, and an overexploitation of ground water is one of the most common signs of this imbalance.

EVER INCREASING WITHDRAWALS FROM NON INCREASING RESOURCES

In some areas, a pincer effect exists between increasing extraction and diminishing resources, but, as a rule, extraction becomes more marked, while the resources fail to grow. In the long term there is a risk that this will lead to a dead end. The problem now, and one that will become ever more pressing in the future, stems less from the disappearance of resources than from an uncontrolled increase in their use.

The need for water is increasing continuously. There are three principal causes of this: the extension of irrigation, population growth and the evolution of lifestyles. Just as in the past, most of the water drawn from nature by people today is used to produce food. During the last century, world agricultural water use increased by five times. In the same period, world population has nearly quadrupled. This demographic boom, along with changes in lifestyle, has brought with it this increase in the demand for water. Modernisation is inevitably accompanied by more intense pressure on resources. Every economic growth point also means one, or perhaps several, growth points in water consumption. In the end, water consumption is growing faster than the population.

----
2. Ghislain de Marsilly, "L'eau, la terre et nous", *Pour la Science*, Dossier "L'eau, attention fragile!" Jan.-Mar. 2008.

1. Water extraction from rivers

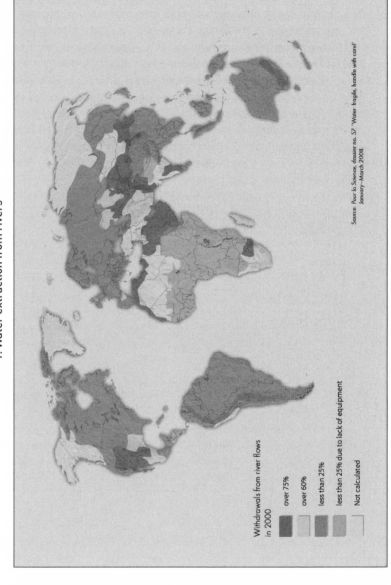

Withdrawals from river flows
in 2000

- over 75%
- over 60%
- less than 25%
- less than 25% due to lack of equipment
- Not calculated

Source: Pour la Science, dossier no. 57, 'Water: fragile, handle with care!'
January–March 2008.

According to the United Nations, fresh water consumption increased six fold between 1900 and 1995, while in the same period the population has merely tripled. If we turn to the future, we can predict that with an annual rate of growth of 1.2 % in 2007, world population will reach 8 billion in 2025 and 9 billion in 2050. To make matters even more complicated, this demographic progression will take place chiefly in countries with a poor infrastructure for water, and in countries where it is already scarce.

For a third of mankind, the future is one of water stress where each such nation will have at its disposal less than 1,700 m³ of water per inhabitant per year. This indicator does not take into account the local and seasonal variability of water resources, but it creates a reference point. Today, 43 countries, i.e. one in five, are already suffering such water stress. The regions where some countries or areas are currently affected are: North Africa, the Middle East, central Asia as far as Iran and western India, sub-Saharan Africa, the south-west of the United States, Mexico, the Pacific coast of Latin America and, to a lesser degree, southern Europe. Year after year, this problem is spreading into new territories. Between now and 2025, it could affect 3 billion people[3]. In North Africa and the Middle East, 90 % of the population may be dealing with water stress. In sub-Saharan Africa the figure is 85 %, and 14 countries may be driven into a state of extreme water scarcity (i.e. with less than 1,000 m³ of water per person per year). According to the Blue Plan, the number of inhabitants of the Mediterranean region who would be "water poor" could rise to 250 million by 2025.

Where water is concerned, many countries and cities or regions are living beyond their means, including the south-western United States, the coastal areas of North Africa, south-eastern Australia, almost the whole of the Arabian peninsula, south-eastern India, and north-eastern China. All these regions of the world are already extracting more than 75 % of the contents of their rivers! If we consider not only rivers but total renewable resources of fresh water, 15 countries, mainly in the Middle East, are consuming more than 100 % of what they possess. Egypt imports more than half of its food because it does not have enough water to produce it itself. China, while it has more water than India, nevertheless faces a difficult situation: a recent Chinese government report warns that "the total volume of water resources exploitable by conventional means will have been reached by 2030, even if efforts are made to economise." With 230 m³ of water per inhabitant per year, where water is concerned Beijing can now be considered as one of the world's poorest cities.

----

3. Human Development Report 2006, *Beyond scarcity: Power, poverty and the global water crisis*, UNDP, 2006.

2. The world's fifty largest conurbations

Source: INSEE/Géopolis, www.citypopulation.de, September 2007; Les Echos, 11 August 2008.

According to the United Nations Environment Programme (UNEP), almost 2 billion people are directly dependent on ground water sources for their water. In a few decades, many American and Asian cities have consumed a large proportion of their underground resources, which were sometimes laid down over centuries. In Spain, 58 % of coastal aquifers are suffering from incursions of sea water. Many communities take more from their underground resources than can be replenished, thus inexorably eating into their capital of fresh water. Even worse, some countries can only survive by exploiting fossil[4] or quasi-fossil resources. Sooner or later these will be irreversibly depleted, certainly within our human timescale. Beneath the vast deserts of North Africa lie colossal reserves of water. Situated under Libya, Tunisia and Algeria, the immense complex of aquifers of the northern Sahara alone contains 31,000 billion m³, that is to say 70 times more than the Albian aquifer of the Parisian basin! It was laid down around 10,000 years ago, when this region enjoyed a wet climate, but more than 2.5 million m³ are extracted each year from these quasi-fossil aquifers, which is more than four times the figure of only 30 years ago. If exploitation continues at this pace, some parts of this gigantic aquifer could disappear within 50 to 100 years.

PARCHED MEGALOPOLISES

In 2008, for the first time in the history of humankind, the number of urban dwellers exceeded that of the rural population[5]. The world of towns and cities has overtaken the world of the countryside. Every month the urban population increases by the equivalent of a city the size of Madrid. In Africa it will double in one generation. The children of tomorrow's Africa, Asia and Latin America will be produced in these rapidly expanding cities. In terms of essential services, the challenges of urban expansion are enormous.

The population of São Paulo is 21 million, while that of Shanghai is approaching 30 million. With 34 million citizens, Tokyo heads the table of megalopolises. Every year, 18 million Chinese leave the countryside in order to settle in urban areas. Even so, it is no longer the megalopolises that are growing the most rapidly, but the middle-sized cities, those with a population of around 500,000. Gigantic urban growth and mushrooming cities pose considerable challenges for local authorities and public service providers. In developing countries, there are few towns

----

4. Fossil watercourses are not replenished. If they are exploited, as is the case in Libya or Algeria, the volume of water contained in them will be reduced inexorably, unless they are refilled artificially.
5. UN Note No. 6144, 22 April 2008.

with their urbanisation under control. In Jakarta, for instance, the water delivery and waste water removal systems were originally designed to serve a population of 500,000. Today, the Indonesian capital has more than 15 million inhabitants and suffers from a permanent shortage of water. The water level, in what was formerly an artesian aquifer[6], is now usually below sea level, in certain places as much as 30 metres below.

Cities in rich countries do not necessarily fare any better. In Riyad, water infrastructure has not kept pace with urban growth and its citizens have running water for only three to four hours a day. At the beginning of the 1980s, the Saudi Arabian capital measured around 20 km². Today it extends over 530 km² – five times the area of Paris – and is expected eventually to reach 2,000 km². With 4.5 million inhabitants in 2008, its population is growing at the impressive rate of 4 % per year.

The city feeds on the blood of the countryside, thought the Flemish poet Emile Verhaeren[7] in the final years of the nineteenth century, as he observed the ravages of urban growth and industrialisation in his native Belgium. We might say nowadays that the city feeds on the water of the countryside. It is certain that many parched cities spill over into the countryside, reaching out with tentacles that capture natural resources. Urban areas are redrawing the map for their own benefit. Cities are "gluttonous": they exacerbate local tensions over water resources and, one way or another, they ensure their own supplies from the surrounding area, with or without its approval.

THE GROWING COMPETITION FOR WATER RESOURCES

Water is not a typical natural resource in that there is a clear distinction between extraction and consumption. Nuclear power stations pump back all the water they have extracted, so they do not consume it, but take it temporarily from rivers. In fact, water that is calculated as consumed, such as water taken by households for washing and cleaning, is also returned in the form of waste water. The only water to be truly consumed is the small fraction that has been drunk or used to water gardens, and even this will sooner or later be returned to the water cycle. At its own pace, making numerous detours, water follows an endless cycle.

Nevertheless, without always realising it, different economic sectors are engaging in a fierce competition to obtain the water they need. On

6. An underground confined aquifer under pressure. In certain conditions, water can even spurt out.
7. Emile Verhaeren, *Les villes tentaculaires*, 1895.

3. Changes in water usage in Africa and Europe

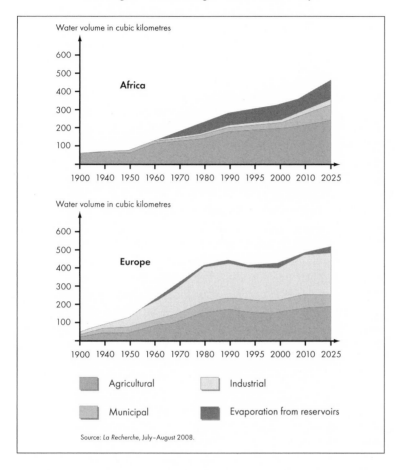

Water volume in cubic kilometres

Africa

Water volume in cubic kilometres

Europe

1900 1940 1950 1960 1970 1980 1990 1995 2000 2010 2025

Agricultural   Industrial

Municipal   Evaporation from reservoirs

Source: *La Recherche*, July–August 2008.

a world scale, the International Water Management Institute (IWMI) estimates that annual water extraction will increase between 2000 and 2025 from 400 km³* to 600 km³ for domestic use, from 800 km³ to 1,200 km³ for industrial use and from 2,600 km³ to 3,200 km³ for agricultural use. In other words, in a quarter of a century domestic and industrial extraction will rise by 50 % and agricultural extraction by nearly 25 %. Even if there are no water shortages for individuals or for industry, there is a risk that there will not be enough for agricultural purposes.

In Paris, the average citizen consumes around 150 litres of water per day for domestic uses; in New York it is 400 litres, in Quebec 800 litres,

---

* 1 km³ equals 1 billion cubic meters.

but a citizen of a developing country consumes only 40 – 50 litres. In Haiti, the figure is 20 litres, and in the shanty towns of Chennai (formerly Madras, in south India) it is 8 litres. In emerging economies, the legitimate aspiration of millions of individuals to access a more comfortable lifestyle will lead to a spectacular increase in household consumption. In Tangiers or Rabat, in Morocco, before being connected to the public network, each inhabitant consumed from 25 to 30 litres of water a day. After connection, the figure reaches 125 litres. We should not, however, condemn this evolution. Modern standards of hygiene involve the use of increased volumes of water. Historically, growth in consumption has gone hand in hand with better hygiene. If it is essential to control the growth in demand for water, it should not be at the expense of public health.

Industry accounts for around 20 % of world water consumption. Over the last century, compared with agriculture and individual users, the industrial sector has shown the greatest growth in the volume of water consumed. Although it is not evident in the final product, water is needed at virtually all stages of production. Between 100 and 120 m^3 of water are needed to produce one tonne of steel, while as much as 11,000 m^3 can be needed to produce one tonne of synthetic fibres. Building cars, extracting and refining oil, all these activities use enormous quantities of water. As Michel Camdessus says, "We drive on water."

WATER FOR AGRICULTURE: THE HIGHEST CONSUMPTION, THE HIGHEST WASTE

World wide, agriculture has higher water consumption than any other sector, accounting for almost two thirds. It also has the highest rate of wastage. In many irrigation systems, more than half the water extracted is lost before it can be used. In a time of crisis over food, the question of agricultural water use becomes crucial, for it determines the capacity of a country to become self-sufficient in food. To produce one kilo of wheat requires 1,500 litres of water, while 5,000 litres are required to produce one kilo of rice. Daily consumption of water by large livestock amounts to between 50 and 100 litres per head. Irrigation is a vital necessity in the majority of developing or emerging countries, and between 1960 and 2000, there has been a doubling of irrigated land world wide.

Agricultural policies, like the "green revolution", which allowed India to become self-sufficient in food, have depended on more productive crop varieties and cheap, plentiful water. But in the long term, these hidden subsidies lead to shortages, because they do not encourage a sensible use of resources. Excessive ground water extraction for irrigation has led to a rapid drop in the main aquifers. Overexploitation of rivers and groundwater comes within the "tragedy of the commons"[8] in which, in the absence of public intervention fixing the conditions and limits of use of a resource, the wealth held in common will inevitably be overexploited.

In 1972, the Yellow River, a giant waterway renowned for its devastating floods, stopped flowing before it reached the sea, having run dry for the first time in its history. In the following quarter century this phenomenon recurred more than twenty times. The periodic drying-up was the result of enormous amounts of water being tapped for agricultural purposes. In 2008, the Yangtze, formerly known as the Blue River, the longest river in Asia, dropped to its lowest level for 144 years. In the same way, widely pressed into service to provide water for agricultural purposes, the Ebro, the Indus and the Rio Grande have also seen reductions in their volume of between 50 % and 90 %[9].

At present, most of the water used for agriculture comes from rain, and not from rivers or groundwater, but future food production will stretch exploitable water resources to their limits. Again, Jean-Marie Fritsch of the DRI explains, "When we look at agricultural productivity and present food production methods, estimates indicate that an additional 5,000 km³ of water will be needed in 2050. This means that for irrigated agriculture nearly four times as much water will have to be pumped from rivers and groundwater sources – which is absolutely unthinkable.[10]" These figures show the extent of the challenge that has to be faced, and yet modern, water-saving irrigation techniques, like drop-by-drop, are used far less in some countries than in others. They are widespread, for example, in Israel, Cyprus and Jordan, but much less so in Tunisia (where they are used on only 11 % of irrigated land), or in Syria (where this technique has been adopted on only 1 % of agricultural land).

Of the three types of water use, domestic, industrial and agricultural, it is the management of the latter that offers the most potential gains. It is largely agriculture that holds the key to the balance or imbalance between mankind and water resources. If the future seems worrying, it will be even more so if we do not prepare for it. And the gap between resources and needs – and the management of this gap – is not a problem only in developing and emerging countries. Everywhere it is

----

8. An economic phenomenon in which competition between several individuals or groups with free access to a limited resource leads to its overexploitation and eventual disappearance.
9. Michel Meybeck, Research Director of the CNRS, "Les fleuves dans le système terre", *Pour la Science*, Dossier "L'eau, attention fragile!", Jan.-Mar. 2008.
10. Jean-Marie Fritsch, *op. cit.*

imperative that we regulate water demands better and arbitrate more fairly between the respective demands of agriculture, industry and households. The famous water crisis is linked less to the lack of availability of water resources than to the way in which they are used. Sooner or later, nature will put a limit on what we can borrow from her, and may exact a savage revenge. Whether or not in anticipation, demographic growth and urban concentration will sooner or later impose a new way of managing water resources.

Water supply depends less on nature than we usually think. Although equatorial Africa has a heavy rainfall, its inhabitants suffer from a lack of water services. Conversely, in the western United States the level of facilities compensates for the lack of water reserves to fulfil the needs of the population. Nevertheless, the growth in demand and the multiplicity of uses mean that there will no longer be enough water for us to have the luxury of using it badly. The age of plenty and lack of concern is coming to an end. The age of bad management of water should also cease.

Total lack of water is the exception rather than the rule and that is the way it should remain. Most countries have sufficient water resources to meet both human and environmental needs. The problem clearly lies in their management. A single image is enough to express it: water losses from leaks in Mexico City are greater than the daily consumption in Rome! The end of "easy resources" calls for greater vigilance over their use. In reality, water is like a great mutual insurance policy: all the inhabitants of a hydrological basin are dependent on each other for better or worse use of their water. These innumerable "mutual societies" of water in Africa, in America, in Asia or in Europe, need to be managed with the utmost rigour in order to avoid what some have called "the exhaustion of nature". If humanity can be water's worst enemy, it is also, when it wants to and when it resolutely gives itself the means, water's best friend. Making water last, while satisfying the needs of mankind, is without doubt one of the greatest challenges to emerge in the twenty-first century.

II. PREVENTING FURTHER LOSS OF THE QUALITY OF OUR WATER RESOURCES

"May man once again become water's friend." With these words Loïc Fauchon, president of the World Water Council, opened the Mexico Forum in 2006. This enmity between mankind and water is recent. It forms part of the wider enmity of modern mankind towards nature. The contamination of rivers and ground water with waste water and agricultural by-products is just one of the most striking manifestations of this wider problem.

Half of the world's rivers and lakes are polluted simply through lack of sanitation. In developing countries, 90 % of waste water is returned untreated to rivers or the sea. In Asia, all the rivers traversing cities are stricken. Only five of the 55 rivers in Europe are considered to be intact. All this pollution not only affects the environment and human health, but also the further availability of water, so that even where water is not scarce, pollution can frequently be so severe that it is impossible to use it. A great variety of products are capable of affecting water quality. We know that water resources can be contaminated by nitrates or pesticides as a result of agricultural practices (spreading of nitrogenous or phosphorous fertilisers, insecticide or herbicide residues, etc.), but above all, the lowering of their quality results from the waste water returned from cities and industries.

SANITATION, THE "POOR RELATION" OF WATER MANAGEMENT

As the year 2008, named "International Year of Sanitation" by the international community, draws to a close, it seems that sanitation too often remains relegated to the background of public activity, and remains the poor relation of water management the world over. At present, 2.6 billion people do not have satisfactory sanitary installations, i.e. private toilets. An even larger proportion of the world's population does not benefit from a system of collection, disposal and treatment of waste water. The situation is even worse because despite the progress that has been made, the number of people without access to sanitation is growing! The right to water must be systematically linked to the right to sanitation, as the latter is fundamental for public health. Providing clean drinking water is an extremely efficient weapon against diarrhoeic illnesses, but establishing sanitary arrangements is even better. The sanitary revolution in Europe in the nineteenth century, which in 50 years added ten years to average life expectancy, was closely linked to the collection and treatment of waste water, as well as to the distribution of clean drinking water. By providing sanitation, we also provide public health. Why, at the beginning of the twenty-first century, do we see such a shortage of access to sanitary provision in the world? The main reason is a lack of urgency from political leaders (whether local, national or international) in putting the problem of waste water at the top of the development agenda. The strong social demand that can be observed here, there and everywhere is rarely taken up by politicians at a high level. How many politicians feel inclined to talk about urine or excrement? In reality, very few. The majority prefer to tackle other more attractive subjects in their public works and action programmes.

The deficit is not only in toilets, but also in the next links in the chain, collection networks and waste water treatment plants. According to the United Nations Environment Programme, in Latin America and the Caribbean 80 % of waste water returned to rivers or the sea does not undergo any treatment whatsoever. In eastern Asia the figure is 90 %. In the areas surrounding the Mediterranean, two thirds of waste water is not treated. In Mediterranean conurbations with more than 100,000 inhabitants, more than half have no waste water treatment plant. As for African cities, most of them do not even have drains. In China, nearly half of water courses and 90 % of surface water resources supplying cities are contaminated and three quarters of its lakes suffer from eutrophication[1]. The Chinese government is aware of the efforts that need to be made with respect to sanitation and has placed environmental protection on its list of priorities. Between 2001 and 2006, as part of the preparations for the Olympic Games, Beijing built six waste water purification plants, and the rate of treatment of waste water in the capital reached 92 % in 2008. Nonetheless, in the most populous country on the planet, transforming this good intention into actual installations in every town and city will take decades.

In megalopolises that have been overtaken by explosions in their population, urban inflation creates unsustainable levels of pollution. The destructive potential of waste water that has been neither collected nor treated is enormous. It becomes "sanitation bombs", to use the striking phrase of Loïc Fauchon, president of the World Water Council, primed, ready to explode one after another. They not only threaten the

---

1. The degradation of water quality due to excess nutrients frequently results in a proliferation of algae.

cities that produced them, but regions further downstream towards which they will be carried.

Urbanisation concentrates large quantities of solid and liquid waste in confined areas. In some districts of cities or in the overpopulated and under-equipped shanty towns in Asia, population density can reach up to 100,000 per km^2! As one inhabitant of an African shanty town said, "Sewage is everywhere." When it is contaminated, water becomes a threat. It destroys health and breeds sickness. Many countries are confronted with vast challenges to bring in sanitation and defuse the "sanitation bomb". Beyond providing private toilet facilities for all, the issue is one of collection and purification of waste water. Many cities in emerging countries have created a minimal sanitation system for their citizens, but for the majority of them, this first phase of individual facilities has not been extended by the construction of infrastructures for the transport and treatment of waste water. In Ho Chi Minh City, family or public toilets have been built by the side of the canals that flow throughout the city. They provide more dignity, comfort and privacy to their users, but from an environmental point of view the waste water situation is the same as before: no treatment takes place and all the waste water is tipped directly into the canals. "The facilities provided merely allow the polluters to be less visible! The ecological problem remains as it was[2]." How many water courses unlucky enough to cross cities, emerge from them transformed into pestilential cesspits?

Other towns have installed long sewage mains, which take the waste away, but without building treatment works to purify the water before restoring it to nature. Finally, when all the component parts of a waste water system exist, namely basic sanitation (private toilets) as well as infrastructures for the collection and cleansing of waste water, it is often the case that they were designed decades ago. As a result of the urban boom, they have reached saturation point and now function badly. For obvious reasons of hygiene, the bigger cities become, the more urgent will be the challenge of sanitation. In certain megalopolises, pollution has reached such high levels that it has become impossible for everyone to be provided with access to clean drinking water without first tackling the problem of waste water treatment. Indeed, many cities are trapped in a vicious circle. On the one hand, urban growth increases the extraction of water from limited resources; on the other, larger quantities of waste water increase the pollution of these very resources. It is possible to escape this spiral but it presupposes good municipal organisation and a political tenacity that will last for many years. Thus western Europe is well on the way to controlling pollution from sewage and industrial

2. Bui Thi Lang and Nicolas Randin, *Métropolisation et pollution à Ho Chi Minh Ville*, EPFL, 1998.

4. Coastal zones threatened with oxygen impoverishment

Human footprint
(Index 0–100)

0–1
1–10
10–20
20–30
30–40
40–60
60–80
80–100

Coastal zones with
oxygen impoverishment

Source: Le Monde, 18 August 2008.

waste. The European Directive on Urban Residual Waste Water Treatment, passed in 1991, considerably improved the collection and treatment of waste water. In France, for example, 20 to 30 years of sustained effort were needed to restore the quality of the water in Lakes Annecy and Du Bourget. In just ten years, from 1998 to 2008, Chile has increased the rate at which it purifies waste water from 16 % to 84 %.

There are many causes of pollution that lower the quality of water resources: direct discharge of waste water into the natural environment, discharges from treatment plants when they are working suboptimally or are overloaded, and excess rainwater which washes through the soil and carries soil pollutants into water courses. In addition to domestic discharges, there is, of course, industrial pollution, whether it comes from active industrial plants or from industrial wasteland, from a continuous process or from an accident, such as the pollution of the Rhine in 1996 after the fire at the Sandoz factory. Newly emerging forms of pollution, such as pharmaceutical residues, present yet another challenge to be faced. In reality, it is essential that we work on all sources of pollution in the aquatic environment. Not one can be considered a lost cause. If even one potential source of pollution is ignored, then all other efforts put into protecting surface or underground water could be reduced to nothing.

We still hear – but less and less, which is an encouraging sign – the argument that a choice must be made between economic growth and protecting the environment. Choosing the first instead of opting for both at once comes down to making only a short-term calculation. In the long term, no country can follow economic development in a degraded environment. If we protect the environment against human pollution, then in return the environment will protect the health of mankind.

A LONG MARCH: CONTROLLING AGRICULTURAL POLLUTION

Although in western Europe pollution from domestic and industrial waste water is under control – or very nearly – the same cannot be said for chronic pollution from agricultural sources. In developing or emerging countries, neither is under control. In almost all countries pesticides and nitrates affect both surface and ground water. The United States is the heaviest user of pesticides, followed by France, which uses 95,000 tonnes of them each year. Europe uses ten million tonnes of fertiliser per year, which is ten times more than in the 1950s. Developing countries today use three times more nitrogenous fertiliser than in 1975. On the island of Barbados, contamination by atrazine, a powerful herbicide, can be as much as three micrograms per litre in wells, which is 50 % more than the threshold recommended by the World Health Organisation (WHO). On Grand Canaria in the Canary Islands, the level of nitrates in water sources has climbed to 170 mg per litre because of banana production, though European regulations demand a maximum concentration of 50 mg per litre in the water supply[3]. Dead marine areas extend along the length of many coasts because of algal blooms, which use up the oxygen. Their surface area is doubling every decade and has now reached 245,000 km², divided into 405 sites[4], including part of the Gulf of Mexico, the estuaries of the Saint Lawrence and Tan Shui in Taiwan, and the Caspian and Baltic Seas. Massive discharges of nitrogen and phosphorus into seawater, mainly from agriculture, have caused these algal blooms, which are asphyxiating coastal waters.

In France, 96 % of surface water monitors and 61 % of ground water monitors show pesticide contamination. As regards nitrates, the quality of water courses has degraded continuously since the 1970s: according to the Institut français de l'environnement (IFEN), the proportion of monitors showing water of mediocre quality (i.e. showing a level of nitrates of between 25 and 50 mg per litre) or poor quality (i.e. more than 50 mg per litre) had risen to 21 % in 2006. In ground water, a similar rise in nitrate levels can be seen. Half of all water courses are classed as medium or poor quality. Over the last few decades, ground water quality in France has continued to deteriorate, while rivers have been improved thanks to heavy investment in cleansing programmes.

In France, as elsewhere, once a problem such as domestic and industrial waste water pollution has been solved, we no longer talk about it. For this reason diffuse water pollution is currently high on the public agenda, and, according to a national survey (carried out by IFOP in 2008) 36 % of the population is concerned about pollution from pesticides and fertilisers.

Sooner or later, a proportion of all fertilisers and pesticides will be found in rivers and water courses. Diffuse and difficult to measure, these pollutants slowly but surely affect the quality of resources. Once they have reached an aquifer, pesticides and nitrates stay there until the water has been replaced. Complete replacement of the water usually takes five to ten years for small underground water tables, but can take centuries for large ones, like the Albian aquifer situated beneath the Paris basin. In many areas, taking into account the quantities of nitrates and pesticides already present in the soil which are migrating towards the aquifers, little by little, year by year, the quality of ground water

---

3. *Groundwater and its susceptibility to degradation*, report of the United Nations Environment Programme (UNEP), 2003.
4. Paper written by Robert Diaz (Virginia Institute of Marine Science, USA) and Rutger Rosenberg (University of Gothenburg, Sweden), published in *Science*, 15 Aug. 2008.

5. Changes in nitrate levels in French water courses

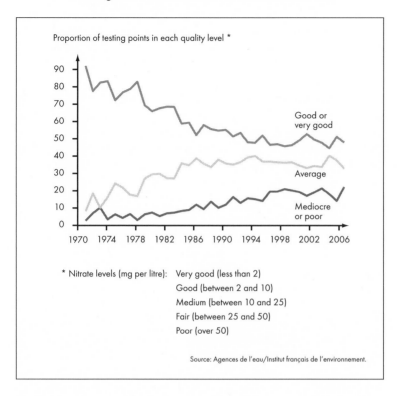

Proportion of testing points in each quality level *

Good or very good

Average

Mediocre or poor

* Nitrate levels (mg per litre): Very good (less than 2)
Good (between 2 and 10)
Medium (between 10 and 25)
Fair (between 25 and 50)
Poor (over 50)

Source: Agences de l'eau/Institut français de l'environnement.

will continue to deteriorate, whatever farmers do. In a way, the dice have already been thrown: even if all pesticide and nitrate use stopped immediately, the figures for underground water pollution would not be turned around for ten years or more.

Public opinion, consumers' associations and the media often blame farmers for nitrate pollution, but while agricultural practices account for the the majority of it, we should not forget that water-treatment plants also discharge nitrates. In the same way, farmers do not bear the entire responsibility for the presence of pesticides in water; they are also heavily used in green urban spaces, along motorways and railway lines. The topic of water pollution tends to revive the old disputes between town and country. Like that of the abstraction of fresh water, it is a source of conflict between these two worlds. Urban water and rural water, city pollution and pollution in the countryside continue to fan the flames of the antagonism between city-dwellers and country folk.

THE LIMITATIONS OF THE "POLLUTER PAYS" PRINCIPLE

One question remains, or rather several, but they are linked and refer back to money: Who should, who can, and who will pay? For waste water as well as for the pollution resulting from agriculture, finance is at the same time a challenge and a bone of contention.

The return on investments in sanitation is not well understood, as the costs due to the absence of waste water collection are diffuse. They are borne by households, businesses and the community as a whole. Unfortunately, it is only when there is a health crisis that we discover the economic benefits of water purification. In 1991 and 1992, the lack of proper systems for drinking water and sanitation caused a terrible cholera epidemic to ravage Peru. It stretched through Latin America, affected a million people and was responsible for 10,000 deaths. In only ten weeks (the time the epidemic lasted), losses from agricultural exports and tourism in Peru totalled a million dollars – that is, more than three times the total amount the country invested in drinking water and sanitation during the 1980s.

In Europe, communities first of all began to tackle industrial pollution, the most offensive, which had the advantage of being localised and not dispersed like agricultural pollution. In addition, it came from businesses which were presumed, rightly or wrongly, to have enough financial resources to pay for cleaning it up, or which, at any rate, were thought to be wealthier than local communities and farmers (which was not wrong). When a town starting from zero – especially in developing countries – must build vast infrastructures for the collection and treatment of waste water, it is impossible to apply the "polluter pays" principle. Sources of finance other than just the users of the water service need to be found, implying a call on both local and national taxpayers, or, if possible, generating public loans over a period long enough to spread the payments on these extremely long-term investments. In France, at the end of the nineteenth and the beginning of the twentieth century, users paid for only half of the cost of purification facilities. The other half was paid by the taxpayer and by the next two generations. In international circles, many speakers stress their allegiance to the "polluter pays" principle, but this is rarely applied completely, even by those communities that have managed to provide the infrastructure. France is one of the countries to have gone furthest in putting it into practice, yet the water authorities there soften the strictness of the principle by adding another under which the "polluter pays" rule is attenuated by a second: "the cleaner-up is helped".

In the case of pollution originating from agriculture, the strict application of the "polluter pays" principle is also unrealistic. Let us remember that it is generally less serious than industrial and domestic pollution, which essentially have both been dealt with in France, mainly thanks to reciprocal aid between users, and public assistance. That is why agricultural pollution appears now to be the most pressing problem in France. Since public bodies, communities and business have succeeded in treating the most dangerous types of pollution, i.e. from factories and cities, they should be able to control agricultural pollution, which is less dangerous. Its diffuse nature, however, makes it more difficult to control. The costs of controlling agricultural pollution cannot be borne entirely by farmers, because they are not able to pass on the costs in the price of their produce, as this is dependent on international markets. Announcing that farmers must pay when they cannot doesn't help to find a realistic solution. Public financial support is therefore necessary, but it needs to be temporary and eco-conditional, i.e. paid only to farmers who make a real commitment to participating in efforts to clean up and prevent pollution.

In Denmark and Germany, local bodies in charge of drinking water distribution allocate financial aid to farmers who agree not to spread pesticides or nitrates in the basins which feed catchment areas. In France, similar systems exist but remain limited. For example, the Ferti-Mieux programme, started in 1992, sets out to modify agricultural practices in the basin fed by the Gorze springs, not far from Metz, where a total of 3,800 hectares is sown with cereals. It has resulted in a lowering of nitrate concentrations below the maximum threshold of 50 mgs per litre. Such initiatives need to be taken up more widely and systematically.

Some consumer associations call for a relentless application of the "polluter pays" principle, which would mean that farmers would bear all the costs of controlling agricultural pollution, but this approach ignores several fundamental points. First of all, as Bernard Barraqué, a water economist and Research Director at the CNRS (Centre national de la recherche scientifique) points out, "If recouping costs, including environmental costs, was done sector by sector, farmers would 'shut up shop' because they couldn't bear the costs of water abstraction *and* the environmental costs at the same time.[5]" Secondly, by what right could users who have not had to finance their treatment plants – which were initially paid for mostly by the taxpayer – demand that farmers alone finance the whole cost of treatment of agricultural pollution? It

5. *La gestion raisonnée de l'eau: vers une autre culture?*, conference of the Cercle français de l'eau, 16 Nov. 2006.

doesn't seem right that users of water services should ask farmers to pay for what they themselves were not capable of undertaking totally, despite being more numerous and, collectively, much wealthier than the farmers. In addition, farmers' use of herbicides is in part a result of market demand for uniform produce. It would not be fair to make farmers entirely responsible for the consequences of consumer behaviour. The final word should go to Bernard Barraqué, who said that when water distribution bodies allocate financial aid to farmers so that they can reduce their need for pesticides and nitrates, "certainly it isn't fair, but it is perhaps better than what might be very fair but would be the most costly for everybody.[6]"

ESCAPING THE SPIRAL: EVER MORE TREATMENT OF WATER THAT IS EVER MORE POLLUTED

In drinking water plants, increased treatment has often become essential, but this solution only treats the symptoms, not the cause. Faced with urban or agricultural pollution, it is vital first to go right to the source so as to conserve the water resources or prevent their degradation, before attempting to restore their quality. The infrastructural heritage of communities, such as systems for providing drinking water and for water purification, must be protected, but the same can be said of their natural heritage, i.e. their water resources. Truly, health security begins with the resource: to protect the resource is to protect the consumer. The old medical adage *"primum non nocere"*, "first, do no harm", can be applied to water as much as to humans as an injunction not to contaminate water, and to clean it up when it has been dirtied by human activity. Moreover, it is easier and more effective to treat a resource that is unpolluted or only a little polluted, than one that is heavily polluted. The more degraded a water resource becomes, the more treatment it requires to make it fit to drink, which results in greater cost to the community and to the consumer. Restoring water quality is good; not degrading it in the first place is better.

The quality of a river is the end result of the quality of the whole drainage basin and the environmental protection infrastructure within it. As the effects of pollution accumulate in them, it is always fresh water courses, as well as coastal waters, that form the real barometer of cleansing policies that have been carried out in the region. Their quality – or lack of it – is the final marker of the general effort made by the community to monitor its waste water.

----

6. *Le bon état de l'eau, regards croisés en Europe*, conference of the Cercle français de l'eau, 17 Oct. 2005.

As much as a century ago, the biologist and writer Jean Rostand warned us, "Man has become too powerful to allow himself to play with evil. His excessive strength condemns him to virtue.[7]" The environment today is affected by many evils, and it is not by chance that it is our short-minded civilisation that has come up with the idea of sustainable development. Too often, mankind has ventured beyond the frontiers of ecological permanence. In many regions, the dysfunctional relationship between mankind and water is proof of this. And yet, faced with pollution of resources and with the many factors that make land fragile, neither mankind nor nature has had its last word. Financial and technical solutions exist, as long as we take account of the realities of land use and show real political will. We know which road leads to become friends again with water; it only remains to follow it with determination. The progressive improvement of water quality begun in Europe proves in any case that failure is not inevitable.

----

7. Jean Rostand, *Inquiétudes d'un biologiste*, Paris, Stock, 1967.

III. EUROPE SETS A GOAL: A RETURN TO GOOD WATER QUALITY

In more than one respect, 2015 will be a crucial year for water management. It is the target date for the United Nations' Millennium Development Goals to be achieved, and it also marks, for the European Union, the end of the period in which member states must restore water quality in compliance with the Water Framework Directive of 2000, a major change of direction in EU water policy.

EUROPE: 30 YEARS OF WATER POLICIES

The construction of Europe goes back nearly 60 years, at least if we take as a starting point the creation of the European Coal and Steel Community (ECSC) in 1951. European water policy is more recent, but it is already crowned with success. The policy goes back 33 years: a third of a century committed to protecting the environment and its inhabitants. One of the most fundamental documents was the 1975 Directive concerning the quality of surface water, followed a year later by Directives on dangerous substances and on bathing water. By protecting the environment, European policies sought first of all to protect Europeans. Several measures resulted from this: the regular strengthening of standards of drinking water quality; the introduction of new limits on substances such as bromates and trihalomethanes in 1998; a regular review of technical advances in analytical monitoring; the transfer of the sampling point, when checking the quality of drinking water, from the public branch-pipe to the consumer's tap. All these key elements are part of a deliberate policy to improve public health, which has been passed on determinedly within the member states by governments and the bodies responsible for drinking water.

In the 1990s the enormous construction work involved with cleaning up water began. The 1991 Framework Directive on urban waste water treatment (transposed into French law by the 1992 law on water) demanded that every city, town and village should have a waste water treatment system, following a timetable according to the size of the community and the date on which they joined the EU (2005 for the first 15 member states, 2015 for the latest 12 to join). This major enactment put water policy at the heart of European environmental thinking. All

WATER

THE WATER CENTURY

over Europe, from capitals populated by millions of citizens to small Alpine towns, people began building, making waste water collection networks and treatment plants reliable and bringing them up to standard. It must be said that some did this more willingly than others, but the movement was launched and there was no turning back. It is mainly through this policy that rivers are no longer "wide polluted avenues": it has restored rivers to greater health and enabled the return of fish species in water courses that had previously been devastated by urban waste. Today, 32 species of fish have been counted in the Seine, compared with only three in 1970. In July 2008, upstream from Paris, a fisherman even caught a sea trout, a migratory fish which demands extremely clean water, and a catch the fisherman had never before experienced.

This water policy – despite having certain weaknesses – was on the whole positive; but it also had its ups and downs. The first difficulty was a result of the creation of what was effectively a new public service, that of waste water cleansing, without explaining to people its ultimate objective, which was to bring large numbers and varieties of fish back into rivers. Nor was it made clear how it was going to be financed by users, even though southern European countries benefited from Cohesion Funds. As a consequence, water bills rose sharply at an economically and socially difficult time, sparking questions about the "fair" price of water. In Spain, a payment strike was organised, spreading into several towns and cities, notably in Catalonia. Elected representatives, administrators and operators seemed collectively to forget that the common good represented by water demands the support of consumers over prices as well as services. They also underestimated the need to educate the public about the new standards, and overestimated the willingness of users of water and cleansing services to pay for the benefits of better protection of rivers, through a sharp increase in their bills.

The second difficulty with this European water policy, and by no means the least, concerned its application in countries with a low population density. Measured by surface area, France is the largest country in the EU, but in terms of density, it is around the European average. Its area (550,000 km²), its population (63 million) and the way it is divided administratively (36,000 communes) combine to produce a multiplicity of sanitation services. It has 17,000 waste water treatment plants, while the Netherlands, with a quarter of the population, has only 450! In proportion to its population, the Netherlands has one tenth as many plants to build, bring up to standard and maintain as France. Bernard Barraqué, Research Director of the CNRS, and a specialist in water policy, thinks that this is one of the reasons why France has been slow to act upon the

Directive on urban waste water treatment. Treatment is, amongst other things, about population density. Geographical differences within Europe also run the risk of differences of approach in the continuing debate over water scarcity. On the one hand, the northern countries are ardent promoters of demand management, especially in the fight against leakages. The southern countries, on the other hand, are more affected by the increasing scarcity of water resources, and tend to call for a supply policy to create new dams and reservoirs.

In the same year, 1991, the European Commission, through its Directive on nitrates, began to tackle the growing problem of the "diffuse" pollution, which is agricultural in origin. Excess nitrates in water (50 mg per litre being the limit for drinking water) can give rise to public health problems in vulnerable people, but above all, nitrates – and also phosphates – cause eutrophisation of coastal waters. Green algal blooms along the coast are the mark of an unsustainable type of development. The challenge posed by the Directive on nitrates for countries where there is intensive agriculture has not yet been picked up. Political procrastination and a lack of clarity are at the root this failure, and have given rise to threats of financial sanctions aimed at several states, including France.

This evolution in European environmental water policy was carried over into the Maastricht Treaty in 1992. Both European water policy and environmental competence already existed. From 1987, the Single European Act had given the EU this competence by confirming the principles of prevention and of "polluter pays". The Maastricht Treaty introduced the idea of environmental policy and placed it among the problems of sustainable development, especially stressing the principle of caution.

This European legislation has profoundly shaped the policies of the member states. Large-scale planning across geographical basins, by means of tools such as – in France – SDAGE (Schémas directeurs d'aménagement et de gestion des eaux (master plans for water development and management)) and SAGE (Schémas d'aménagement et de gestion des eaux) (plans for water development and management)) or their equivalent in other countries, turned out to be an asset for concerted action. The European Framework Directive of 2000 placed a requirement on all member states to organise themselves by hydrographic basins. This acknowledged the relevance of the French method of organising by basins, which had begun in 1964. Unfortunately, in doing this, it is all too easy to forget the factor of finance and the principles of mutual solidarity between the players who were at the heart of the French law of 1964, which had created what were then known as the Agences financières de bassin.

GOOD WATER STATUS: MOVING TOWARDS A NEW STAGE IN EUROPEAN WATER POLICY

The Framework Directive of 2000 was a turning point in European water policy, the full consequences of which have still not been completely felt, even today. It effectively replaced an environmental policy based on control of use and disposal with an approach centred on the quality of the environment. In stating that Community waters must achieve a "good" status by 2015, the EU has set itself a Sisyphean task, namely to regain control of and to preserve water resources. This will no doubt involve work on a scale as enormous as the cleaning up work decided on in 1991, which has still not been completely finished. In considering things "from the river's point of view", the 2000 Directive encapsulates new ambitions: to gain control of pollution at its source, to put water at the centre of a new approach to the protection of living things and of sustainable development, and to respect the right of future generations to a high-quality environment. Rivers should not be appraised using only physical and chemical parameters, but also using biological criteria: chiefly, the presence, nature and number of invertebrates, diatoms and fish species. This brought Brice Lalonde, the former French Minister for the Environment, to say: "one could almost say that European water policy is aimed at satisfying the needs of aquatic macro invertebrates: insect larvae, molluscs, leeches and worms[1]."

Even beyond its object, the Framework Directive of 2000 is revolutionising our approach to European politics. First of all, it imposes on member states the objective of a result, not simply a means, alongside a strict deadline: 2015 or 2025, according to the area and the country (it is 2025 for new entrants to the EU). At the same time, the principle of subsidiarity demands that a certain degree of flexibility should be included in the definition of the final "good status" that is required, as well as in the methods for covering the costs. The member states have room for manoeuvre in defining "good" and "bad" as regards the quality of water masses, and each state has to work within its own administrative and technical traditions, as well as within the demands of its own environment. The European Commission wanted to take account of this, and it has led to friction between more and less demanding countries, to incomprehension because member states were asked to achieve an undefined objective, and to foot-dragging. We now see that half of the 15 years assigned by this Directive have been devoted to defining the nature of "good", so it is clearly true that member states were asked to

----

1. *Le bon état des eaux, regards croisées en Europe*, colloquium of the Cercle français de l'eau, 17 Oct. 2005.

6. The 2000 European framework directive on water: a simplification of the law

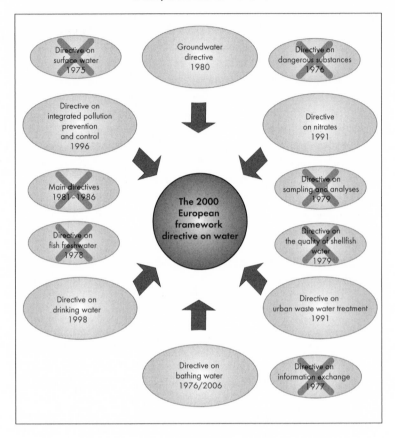

achieve an undefined objective! But it remains resolved that all EU citizens must have good quality water resources by 2025 at the latest. This progressive approach, which gives each member state some latitude, is a mark of pragmatism, and it respects the starting point and the pace of change appropriate to each country.

Next, the 2000 Directive introduced a financial element by asking that the expenditure to be undertaken to achieve these objectives should be assessed according to the means available. This responded to the lack of economic planning in many member states. Finally, it gives more public participation in the decision-making process and in consultation, for example, over the restructuring of the SDAGE and SAGE in France, or of their equivalents in other European countries.

In addition, the Framework Directive reinforces and at the same time simplifies legislation, thanks to it integrated approach to problems. It is not simply a supplementary set of regulations, but a major text reworking some pre-existing pieces of legislation and cancelling others: in fact it has led to the repeal of seven previous Directives.

A DISAPPOINTMENT ON CURRENT WATER QUALITY

The report on the principle water masses of Europe, published in March 2007, reveals a worse situation than expected, in that 40 % of them will not reach "good" status by 2015. The deterioration there has been in water quality is due to diffuse pollution and physical changes, but also to the overexploitation of resources, especially underground. Only 30 % of water masses are not classed as "at risk". Among the other water masses there is a category for which not enough information is available for an assessment to be made. It should be noted, however, that a great many problems have already been solved in states that have applied what are called the guideline Directives in their entirety. This is not the case in countries that have been more negligent in putting these laws into practice on the ground, nor in the new EU member states.

Faced with a state of affairs which makes working to its timetable simply hypothetical, the European Commission issued three recommendations: to speed up the implementation of the Directives on urban waste water treatment and on nitrates; to put in place a co-ordinated system of information; and to integrate sustainable management of water into other policies, especially agricultural and regional policies. It is consistency between these different European policies that will ultimately achieve the desired "good" status.

QUALITY AND QUANTITY: FACTORS SEPARATED
FOR TOO LONG

The European Union was late in revealing the problem of water scarcity. It began to worry about it more seriously after the drafting of the 2000 Framework Directive, which requires member states to ensure good water quality as well as good quantitative status. Nevertheless, the recurring droughts which affect large areas of Europe make this an even more pressing issue. In fact, quality and quantity are inseparable. Droughts affect the ability of water courses to clean and purify themselves naturally. They also disturb ecological continuity and the relationship between river beds, marshes and wet zones. They damage

7. Compliance with the 2000 European Framework Directive by EU Countries

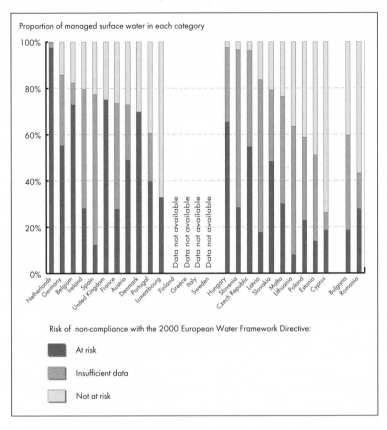

Proportion of managed surface water in each category

Risk of non-compliance with the 2000 European Water Framework Directive:

■ At risk

▨ Insufficient data

▫ Not at risk

spawning grounds[2] and compromise fish reproduction. In the summer of 2003, at Chinon, water temperature in the Loire exceeded 33° C! The succession of local droughts and the problems that accompany them (agricultural losses, public health problems, conflicts over the use of remaining water resources, restriction of industrial activity, etc.) could become even longer. It spurs us to move more quickly towards global management of resources in terms of both quality and quantity.

According to the European Commission, 11 % of the population of Europe and 17 % of its surface area are affected by water shortages. The bill for droughts over the last 30 years has been estimated at 85 billion euros. The drought of 2003 affected 100 million inhabitants in a third of the EU area. Faced with the shortages that are taking shape, the Commission's proposed response comes down essentially to a policy of

----

2. Marine locations where fish gather to breed.

restricting the use of water and preventing it being wasted. This response has been judged insufficient by many Members of the European Parliament (MEPs), who remind people that while the Commission may put forward initiatives and propositions on water policy, European legislation on water is ultimately adopted by the European Parliament and by the Council. It is possible that in future MEPs will make their voices heard more clearly on this issue, especially as other European policies which impact on water policy (like the Common Agricultural Policy or regional policy) are the exclusive responsibility of the Council. It is likely, and it is certainly desirable, that in resolving this issue geography should take precedence over national affiliations.

IV. WATER: THE PRIMARY ISSUE FOR HUMAN DEVELOPMENT

Following the previous chapters where the growing needs for water and the fight against resources pollution were highlighted, let's come to the third and most urgent challenge: access to safe drinking water and basic sanitation. In 2008, we are not necessarily aware of what running drinking water represents in terms of public health and comfort. We have forgotten the decisive progress made by the revolution in hygiene in the nineteenth century, and by the introduction of piped water into towns and cities. And yet, all over the world, whole populations are denied access to safe drinking water and drink dirty water instead. Today nearly one billion people do not have access to clean water, and 2.6 billion lack basic sanitation, i.e. hygienic private toilets.

Reducing this proportion by half between now and 2015 is one of the eight commitments made by the United Nations in its Millennium Development Goals. In September 2000, world leaders declared their wish in the medium term to halve the number of people lacking access to clean water. Then, after the World Summit on Sustainable Development in 2002, a supplementary goal was added: to halve the number of people without access to basic sanitation between now and 2015. Taking account of demographic growth, reaching the Millennium Goals means supplying safe drinking water to 900 million people and basic sanitation to 1.3 billion before the target date.

Access to water and sanitation is not merely one Millennium Goal among others. It is the most important, because without it the other goals, the struggle against poverty and hunger, reduction of infant and maternal mortality, protection of the environment, universal elementary education, promotion of the role of women, etc., cannot be achieved. Water is essential for an acceptable way of life, and without it there can be no health, no education, no development. The issues around water affect every aspect of life.

UNSAFE DRINKING WATER AND NO SANITATION: THE WORLD'S GREATEST KILLERS

"We drink 90 % of our illnesses," Louis Pasteur used to say. The absence of safe drinking water and infrastructure for sanitation, and

8. Access to installed water supply and sanitation

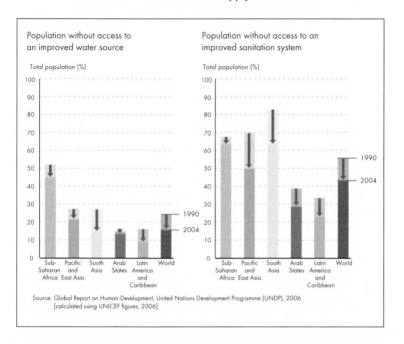

Source: Global Report on Human Development, United Nations Development Programme (UNDP), 2006 [calculated using UNICEF figures, 2006].

the water-borne illnesses which result from this, are the greatest cause of mortality and morbidity in the world – above even malnutrition. It is estimated that half of all hospital beds in the world are occupied by patients suffering from illnesses directly or indirectly linked to water. Intestinal worms affect 10 % of the population in developing countries, six million people have been stricken with blindness caused by trachoma, 90 cholera epidemics have been declared since 1996, and 200 million people are infected with bilharzia. Malaria and dengue fever, transmitted by larvae present in stagnant water, are among the world's greatest epidemics, in terms of numbers of people affected. Each year, 1.8 million children die from diarrhoea, which means 5,000 children a day. That is more than the total of all the wars in the decade beginning in 1990. According to the 2008 WHO report *Safer Water, Better Health,* diarrhoeic diseases cause 100 times more deaths in developing countries than in developed countries. For Kofi Annan, the cause is clear. In 2001, when he was Secretary General of the United Nations, he declared, "We shall not defeat AIDS, tuberculosis, malaria or any other of the infectious diseases that plague Africa until we have also won the battle for basic health care, safe drinking water and sanitation[1]."

----

1. Kofi Annan, Durban Conference, 30 July 2001.

Public health is one of the issues that have historically been linked to the development of public water and sanitation services, and remains so in developing countries. At the end of the 1890s, infant mortality in Great Britain was comparable to that in Nigeria today. In the twenty-first century (as in the nineteenth), only a strong political will will allow these problems to be resolved. Let us remember the great cholera epidemics that struck Paris in 1832, Marseilles in 1835 and London in 1832 and 1849. In France, piped water was rare before the middle of the nineteenth century. Water carriers, who drew their supplies from the Seine upstream of the city, were found all over the streets of Paris. Change began in the capital with Baron Haussmann, Prefect of the Seine area from 1853 to 1870, who had a drinking water system built, drawing water from sources upstream of the Seine, and who modernised the sewers of Paris. The UNDP stresses strongly that "water and sanitation are among the most powerful preventive medicines available to governments to reduce infectious disease[2]."

The issue of sanitation is no less important than that of access to safe drinking water. As the old African proverb puts it, "Water cleans everything, but it is very difficult to clean water." The lack of sanitation and basic hygiene is an everyday problem for almost half the world's population. As just one example of many megalopolises in emerging countries, two worlds live side by side in Mumbai. The city has mobile phone and internet networks, while cuts in water and electricity supplies are a daily occurrence, and sewers are open to the skies. Office rents in the business district, Nariman Point, are as high as in the City of London or the centre of Tokyo. Just a few metres from this Manhattan of Mumbai, children dive into filthy water all along beaches strewn with rubbish. In the slums of the Indian economic capital, as in those of the rest of Asia, Latin America or Africa, thousands of people have mobile phones, but they still do not have safe drinking water.

In Kibera, a slum district of Nairobi in Kenya, and the second most populous slum in Africa, latrines are few and far between. After nightfall, around 7 p.m., it is dangerous to leave the house, so people defecate into plastic bags which they throw into the street. They call these "flying toilets". In this slum, the infant mortality rate is almost double the national average for Kenya. Throughout the world, people are forced to relieve themselves in bags, buckets, fields or ditches at the side of the road, which carries great risk to their health and that of others. In private houses or in schools, there are frequently no toilets, nor is there any soap for hand-washing. Children are more likely than adults to touch dirty surfaces and are thus especially vulnerable. In addition, even when

WATER

THE WATER CENTURY

2. Human Development Report 2006, UNDP, *op. cit.*

toilets are installed where there is no system for evacuation or treatment of waste water, there are many potential sources of contamination.

Drinking water taps and toilets, the waste water collection systems that revolutionised public health in New York, Paris and London a century ago, are unfortunately too little used in the fight against poverty and disease in developing countries. Nevertheless, just as they always did, these tools function brilliantly from the moment they are put into use. In Uganda, better access to safe drinking water has reduced the infant mortality rate by 23 %. Access to water-flushing toilets has cut the infant mortality rate by 59 % in Peru and 57 % in Egypt when compared with households without access to such basic sanitation facilities[3].

LACK OF ACCESS MEANS A HIGHER PRICE FOR WATER

For people who are already at a disadvantage through being uncon- nected to a water supply, it also implies considerable extra expense. An inhabitant of an African slum who is not connected to the public system must get his water supplies via one of the parallel systems, which will charge up to twenty times as much for a litre of water as any normal col- lective system. A large proportion of the population in developing coun- tries buys its water from resellers who sell at a much higher price than that paid for piped water. Slum dwellers in such cities as Jakarta in Indonesia, Manilla in the Philippines or Nairobi in Kenya, pay between five and ten times more for their water than people in more prosperous districts of their city, and in Baranquilla in Colombia, they pay up to ten times as much as a citizen of New York. In Ho Chi Minh City a few years ago, the farther one travelled from the central districts to the farthest perimeter of the city, the more the price of water increased, with people not connected to the supply having to pay between three and twenty times as much per litre as those connected to the public supply.

Very often in Africa, water is bought from the manager of a commu- nity-owned stand-pipe. The customers pay either with money, or with a token that they can buy in advance. Twenty-litre jerrycans will also be on sale nearby. Stand-pipe managers, water carriers, water resellers, water truckers with tanker lorries, all these small operators in the popular economy play a key role in the water sector. On the African continent, they supply half of all city-dwellers; in Latin America and east Asia, a quarter. Although they play a vital role for millions of people, they usually operate within a legal no man's land, being neither completely legal, nor explicitly illegal, but they sell water at a higher price than from any

----
3. Human Development Report 2006, UNDP, *op. cit.*

public system. In Africa and Latin America, the price per litre from a stand-pipe is on average double the price from the public system, and in south Asia it is five times more expensive[4].

9. Water prices by forms of supply

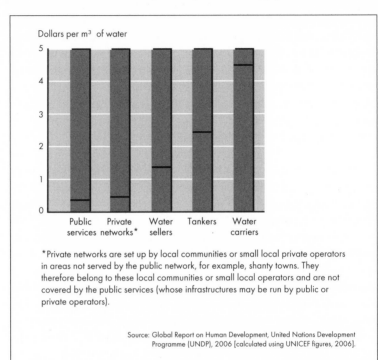

Dollars per m³ of water

*Private networks are set up by local communities or small local private operators in areas not served by the public network, for example, shanty towns. They therefore belong to these local communities or small local operators and are not covered by the public services (whose infrastructures may be run by public or private operators).

Source: Global Report on Human Development, United Nations Development Programme (UNDP), 2006 [calculated using UNICEF figures, 2006].

Not being connected to the public drinking water system also means that customers receive water of a lower quality. Even when water is provided from boreholes, wells or stand-pipes, it is generally of a reduced quality because of inadequate maintenance of the equipment. The containers used to draw water from wells are often placed directly on the ground after use, where passing animals may drink from them. The large number of intermediaries involved in transporting the water contribute to making it unfit for consumption. It can be stored in a courtyard or even in a corner inside a house. In large African cities, so-called "pure" water (which is often not fit to drink) is sold at crossroads in plastic sachets. More than water itself, it is often information that is needed here.

----

4. From the study published in 2005 by Kariuki and Schwarz, which covered 47 countries and 93 geographical locations.

As a consequence of all this, families that are not connected consume much less water than those that are, and often a lot less than good hygiene demands. In Ulan Bator in Mongolia, people who live in the yurt district get their water supplies from kiosks and consume on average only 5.7 litres per person per day, 50 times less than those who live in the city centre where there is a public supply, and four times less than the minimum of 20 litres per person per day recommended by the WHO.

Unequal access to sanitation is compounded by unfair prices. Poor access to water leads to an absurd water economy, where deprived people pay much more for their water than the affluent. Even worse, the poorest in developing countries sometimes pay more per litre than people in developed countries. Social justice, on a local as much as on an international level, calls for a rationalisation of the water economy.

WASTED HUMAN RESOURCES, WASTED SOCIAL RESOURCES

Water-related diseases lead to absences from work and force children to miss school. It is estimated that in total 443 million school days are lost every year[5]. Girls in particular, especially when they reach puberty, cannot go to school for lack of clean, secure latrines. In fact, many parents forbid them to go there because the schools do not have these basic facilities. For these children, the right to education depends on the right to water and sanitation. This has been demonstrated by activities specifically targeted at educational facilities. In Tangiers, UNICEF, AMENDIS (a subsidiary of Veolia), Veolia Waterforce and the Ministry of National Education led a programme to equip 31 schools which had no sanitation, water or electricity. It affected the lives of nearly 40,000 pupils and cut school absences down to one eighth of what they had been.

Water lies at the heart of the injustices done to women. In Africa, women often work four or five hours a day more than men, and every year they carry, according to where they live, tens of tonnes of water and firewood. The distance they must travel in Africa and Asia in order to provide water supplies is sometimes as much as 10 km. The necessity to carry water over long distances and to stand in long queues before they can get their water means that they lose time and energy. They return from these daily chores with illnesses and injuries. According to the WHO, the lack of universal access to safe drinking water and sanitation means that sub-Saharan Africa loses 5 % of its GDP every year,

5. Human Development Report 2006, UNDP, *op. cit.*

which amounts to more than the total of international aid granted to the region. The World Bank estimates this loss at 9 % in Pakistan and Ghana.

But figures can never express what lack of water means. The issue is not simply one of individual health nor, for economists, of a country's GDP. More intangible, but no less important, it is also one of human dignity. In the end, for poor families, not having access to drinking water and sanitation means extra cost, a sub-standard service, more illness and less school.

MILLENNIUM GOALS: MIXED RESULTS IN THE MEDIUM TERM

Is it appropriate to be pessimistic about the great hopes born of the Millennium Goals? Even if, once commitments have been made, action on the ground has started up more slowly than expected, even if projects fail, this is no reason to resign oneself to failure. Mobilisation brings its own rewards. With all that said, however, the efforts being made are still insufficient, particularly for sanitation.

Already many countries have made notable progress. In Uganda, in ten years, five million people have gained access to clean water. This country has increased its water and sanitation budgets from 0.5 % to 2.8 % of public expenditure. It is not insignificant to point out that Ugandan legislation requires that women be represented at the centre of all water-users' associations. In South Africa, since 1994, ten million more people have gained access to clean water. This is one of the few countries in which public expenditure on water and sanitation exceeds the military budget. Ten years ago, in Morocco, very few people in the countryside had safe drinking water, but in the last decade four million more people have gained access to drinking-water systems in the countryside, which brings the total for rural areas to 50 %. At the same time, rural primary school attendance for girls has leapt from 30 % to more than 50 %. In Niger, an institutional reform accelerating the progressive transfer of responsibilities to local communities and the private sector has increased the number of branch-pipes. In Gabon, the number of people connected to a modern drinking-water system has nearly doubled and the country has met its Millennium targets ten years ahead of schedule.

There have also been successes in the field of sanitation, such as in West Bengal, where spectacular advances have been made. In 1990, when the government of this Indian state launched its rural sanitation offensive – the largest in India – in the district of Midnapur the rate of coverage in rural areas was no higher than 5 %. It is now at 100 %. Over the whole state of West Bengal, two million toilets have been built in the last five years. The level of provision of sanitation services has now topped the 40 % mark. Thailand recently reached 100 % nationally in sanitation coverage, having made remarkable progress in rural areas, where more than 13 million people have been connected in the last two decades[6]. Within the framework of a national strategy, each Thai district was invited to identify the gaps existing in its system and to devise a plan of action to remedy them.

In its global Human Development Report, published in 2006 but referring to statistics for 2004, the UNDP makes an intermediate assessment of the Millennium Goals. It emerges that, give or take a year, the world is very close to reaching the target as regards drinking water – due particularly to the progress made by China and India. Today, based on the figures for the end of 2006, the WHO and UNICEF estimate that no more than 884 million people still do not have even minimal access to safe drinking water, compared with 1.1 billion at the beginning of the 2000s. This recent data leads us to believe that globally, the Millennium Goal relating to drinking water will be exceeded in 2015.

However, this global assessment hides profound differences. If the pace of change is not increased, many countries will fall short of the Millennium Goals and, in 2015, hundreds of millions of people will still not have access to safe drinking water. Sub-Saharan Africa will reach the goal a generation after the due date. Today, 38 % of the population of Africa and 19 % of that of Asia still do not have access to clean water. For Latin America and the Caribbean, the figure is 15 %. Also, in some parts of the world, demographic growth wipes out any progress made in providing facilities. In sub-Saharan Africa, despite much work having been done, the number of inhabitants without access to safe drinking water has increased by 60 million since 1990.

In addition, the rural world is often forgotten in water policies and there are huge disparities between urban and rural areas. If we look at developing countries, drinking water provision is as much as 92 % in urban areas, but in rural areas it is 20 percentage points lower. The gap widens for sanitation: provision in rural areas is less than half of that in urban areas. The greatest disparity between cities and countryside in access to essential services is in sub-Saharan Africa, where rural inhabitants are four times less likely than city-dwellers to have access to basic sanitation. There is no surprise here: cities are centres of wealth and have greater financial means than the countryside to build infrastructures. In India, for example, cities produce two thirds of the GNP.

----

6. Human Development Report, 2006, UNDP, *op. cit.*

If, globally, it seems possible to reach the Millennium Goals for drinking water, the same cannot be said for sanitation, which is a long way behind in the majority of developing countries. Even though 2008 was nominated "Year of Sanitation" by the UN, the situation is far from ideal. Rates of provision are extremely low: only one person in three in sub-Saharan Africa and south Asia has access to basic sanitation; in east Asia it is one in two; in Jakarta and Manilla, one in ten. Although global provision of basic sanitation increased from 54 to 62 % between 1990 and 2006, we are still a long way off the target and there is a real risk that it will take two generations to achieve the Millennium Goals! Seventy-four countries are behind in the race and 700 million people will miss out in 2015 as far as the agreed objective is concerned. These delays are even more serious in that millions of people have been left out of the statistics, which cannot then show the full extent of the deficit.

To make things more difficult, when the Millennium Goals were agreed, the United Nations did not define precisely the meaning of access to "basic" sanitation. While some organisations use the definition "hygienic private toilets", and even though work is in progress to clarify the concept, nobody really knows what it means, and people are still free to interpret it as they please. This is to say nothing of the fact that, beyond basic sanitation, there is a huge gulf between simple communal or private toilets and a true sanitation system which will take waste water far away from areas where people live; between a sanitation system which discharges untreated waste water into rivers and another system with a treatment plant that can purify the water before returning it to the environment. People who live in developing countries are waiting for a lot more than the Millennium Goals! They do not wish only for better wells, but for safe drinking water systems and taps to give them running water at home. They do not just hope for private toilets, but ask that they should be protected from contamination by their neighbours' effluent, which means collecting and disposing of waste water.

These gaps in terms of access to essential services have repercussions on the public health figures. The high level of mortality resulting from a lack of safe drinking water and sanitation marks Africa out from other continents. According to the 2008 WHO report *Safer water, better health*, mortality due to lack of access to safe drinking water and sanitation stands at 24.1 % in Angola, 23 % in Niger, 20.9 % in Mali, 20.4 % in the Congo, 19.7 % in Madagascar, 19 % in Benin and Burkina Faso, 18.5 % in Chad, 17.8 % in Liberia, 17.7 % in Mauritania and 15 % in Ethiopia. Outside Africa, countries where the figure is above 15 % are the exceptions, but they include Afghanistan, with a rate of 16.2 % and Yemen, with 16 %.

To make up for lost time, sub-Saharan Africa will have to double its "cruising speed", going from 10 million to 23 million people per year connected to drinking water systems. It will have to quadruple the rate at which it expands sanitation provision, connecting 28 million people per year between now and 2015, instead of the present 7 million. South Asia will have to provide access to a sanitation system to 43 million people per year, compared to the 25 million people who have been connected annually during the last decade.

10. UNDP forecasts for reaching Millennium Development Goals

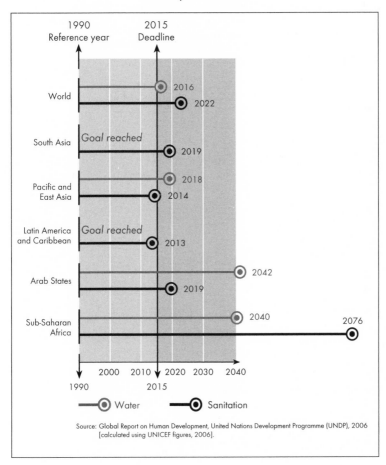

Source: Global Report on Human Development, United Nations Development Programme (UNDP), 2006 [calculated using UNICEF figures, 2006].

PART 2

WATER'S FALSE FRIENDS

Although the global situation is urgent, it is far from being desperate. In order to succeed we need everyone to have a sharper sense of responsibility, more effort in research and innovation, stronger governance, finance, and a commitment to keep commitments. But we must also rid water of a pernicious plague: pointless quarrels. The world of water professionals seems to get worked up by virulent polemics, as if to avoid assuming certain responsibilities. Let us be clear: water is a political subject *par excellence*, so it is quite normal that questions over water should be riddled with debates, even antagonisms. But is that a reason to resort to extreme over-simplification or unthinking ideology? Too many arguments make us lose precious time; neither water access nor water quality will be improved by invective.

It is time to put an end to various false debates. They divert attention from questions that really do need more discussion and they conceal part of the reality in order to justify the *status quo*; they act as pretexts for impotence and defeatism.

I. CLIMATE CHANGE AND WATER: DISTURBING FACTS

UNAVOIDABLE CHANGE, AVOIDABLE CATASTROPHE

Over the years, the alerts surrounding climate change initially raised by experts have spread to inform public opinion. Summer heat waves, frequent floods, the publicity for Al Gore's film *(An Inconvenient Truth)* or quoting the environment as a theme in election campaigns have all helped this new awareness.

The reality of climate change is incontestable. The average temperature of the planet surface has increased by 0.75° C over the last 100 years. France has grown hotter by 1.4° C since the beginning of the twentieth century, with rapid acceleration over the last three decades; Switzerland by 1.5° C; Canada by 1° C. In Nepal, average temperatures have increased by 1° C just since 1970. This is without mentioning the retreat of the Arctic ice cap, the delays in snow coverage in the mountains, the early Spring.

The 2001 report of the Intergovernmental Panel on Climate Change (IPCC) proves that the essential cause of this warming is greenhouse-effect gases left in the atmosphere by human activity. Water is at the heart of the latest IPCC report, of 2007, and its conclusions are stark for the planet's future. By 2080, 3.2 billion people will suffer severe lack of water, and 600 million will suffer from famine caused by drought and soil degradation. It forecasts a decrease in water resources across much of Asia. At the same time population growth and higher living standards will require ever increasing quantities of water. In Europe, nearly all regions will be negatively affected: floods, erosion, ecosystems, thawing of glaciers and the disappearance from the continent of 60 % of species between now and 2080. Southern Europe will be the worst affected: there will be a reduction in water provision and hydroelectric potential, an impact on tourism, lowered agricultural productivity.

Although climate change is certain, its effects are uncertain. It would be presumptuous to claim to know the effects of global warming on water resources. There are predictions, but they differ by offering contrasting scenarios. For example, it is estimated that climate change in Africa will endanger water provision for between 75 million and 200 million people between now and 2020. But there is a considerable difference between 75 million and 200 million, and adaptive measures – which will be necessary whatever happens – will vary greatly. However, one thing is certain: climate change will exacerbate tensions, because it will make water scarcer in many places. So conflicts over the use of water risk becoming fiercer – and these conflicts will mainly be between priorities. The outcomes resulting from these conflicts may well be disastrous for some users.

IPCC experts regard water resource problems as a major effect of global warming. Schematically, they predict that it will rain more in regions that are humid today (tropical regions) and less in arid regions. Climate changes over recent decades have already exacerbated the problems of some regions. Over the last 25 years, the Sahel has seen the most sustained decrease in rainfall ever registered in the world, punctuated by recurrent droughts in Burkina Faso, Mali and Niger. In West Africa, river flows have fallen by more than 40 % since the 1970s[1]. Lake Chad is disappearing. The main responsibility for this is man-made – an increase in water drawn from the lake – but rainfall reduction has also played a considerable part.

When we take into account the increasing future demand for water and the limited nature of easily available resources, climate change will clearly erode living standards in poor populations. Environmental degradation and desertification may drive some of them into exile. Recently, we have already seen this happening. Between 1960 and 1980, droughts and desertification in north-eastern Brazil led to the displacement of 3.4 million people[2]. Unfortunately, such tragedies will recur. Among future environmental refugees there will now be "water refugees". A study carried out in preparation for the Stern report[3] estimates that a minimum of 146 million people will be made directly vulnerable by rising sea levels.

Nearly everywhere in the world, climate change leads to the retreat of glaciers. Over the next few decades, the ice thawing in the Himalayas will, at first, increase flooding. Later on, glacier retreat will lead, lower down, to the drying-up of rivers and a reduction in available surface resources. Water abundance will be followed by water scarcity. The results of this glacier thaw on the availability of water will affect 500 million people in the Ganges valley, 250 million in China and 10 million in the Andes[4]. By storing less water in the form of ice, glaciers will release less water during the dry season to maintain river flows.

----

1. Source: 2006 Global Human Development Report, UNDP, *op. cit.*
2. Étienne Piguet, "Migrations et changements climatiques", *Futuribles*, no. 341, May 2008.
3. *Review on the Economics of Climate Change*, by Nicholas Stern for the UK government, published in 2006.
4. Stern, *op cit.*

11. The disappearance of Lake Chad

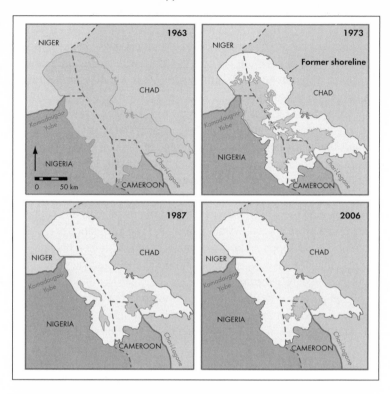

Climate change displaces rainfall and makes it more intense. In areas that receive more rainfall, water availability will not necessarily increase to the same extent. It might even decrease. In fact, groundwater is not replenished much by violent rainfall: most of the water falling from the sky runs off over the soil without soaking into it, and then flows into water courses. If the water runs away too quickly, it cannot be absorbed, and it flows away without being usable, unless vast dams are built to hold it back. So, in countries that will receive more rainfall, the real availability of water will depend, not so much on the annual rainfall figure as on its distribution over time and the speed of its circulation. For any given quantity of rainfall, the final amount of water available will vary according to whether it occurs as storms or as gentle rain spread throughout the year. The effects of climate phenomena on water resources do not occur at the same rhythm or over the same area as the effects of increased withdrawals for water supplies. The former are measured over decades for whole sub-continents, whereas the effect of the latter can be observed rapidly, over a few years or less, and on a localised scale.

Global warming is a plausible explanation for various violent climatic events we have seen since the 1990s. The number of extreme weather episodes is likely to increase over the coming decades. Unfortunately, the cyclone map largely corresponds with the map of the most populated areas, with a growing percentage of the world population settling on the coast. However, we need to assess developments by setting them in a long-term context: the rising waters of the Elbe seen in 2006, for example, which caused so much comment, were lower than those that occurred in the nineteenth century.

With the media continually haranguing us about global warming, we tend to forget that our current water problems in the world derive from human activity, and not from nature. They are caused mainly by population increase and the increased demand for water per person, rather than by climate changes. In developing countries, access to drinking water for all is a question of political priority and not one of the gross availability of water resources.

THE CLIMATE CHANGES... AND US?

For Jean-Marie Fristch, a hydrologist and the IRD (International Relief and Development) representative in South Africa, "the water crisis will not take place. People are getting more and more worried about a coming global water scarcity. However, the problem is not inevitable, and has a lot less to do with the quantity of water than with our manner of using it. [5]" In other words the outcome lies in the hands of water users. Although climate change may be unavoidable, catastrophe is not inevitable. The extent of its ill effects will depend on our collective capacity to bring forward policies designed to attenuate its effects (policies designed to limit greenhouse gas emissions) and policies to adapt to a changed climate. Of course, we cannot master the climate but we can manage the water.

The problem of climate change, which overlaps today with the financial crisis and the ecological crisis, demands a different response from the pervading catastrophism or fevered activism. Uncertainties, not about the reality of climate change but about how severe it will be and its local consequences for water, mean that we must identify what we need to change in order to adapt. Responses to the questions raised by climate change do not depend on science alone, but on the will to put collective values into practice: fairness between generations, an

----
5. Jean-Marie Fristch, *op. cit.*

65

equilibrium between nature and human activities, solidarity between countries, trust between those involved.

FORESEEING TO BE ABLE TO ACT

"Do not foresee, and you start groaning," said Leonardo da Vinci. Denying the reality of climate change and its effects on water availability would be both absurd and backward. Such a denial would be a rerun, for instance, of the mistakes committed in Brittany on the role of nitrates and pesticides in the degradation of natural resources. Refusing to face facts ultimately leads to an interminable catch-up race to try to regain the time lost, and, after the event, correct evils that could have been avoided.

In the face of climatic and hydrological change, it is vital to place scientific knowledge at the service of practical action, to build working scenarios to deal with probable futures. This is the strategy adopted by the Seine-Normandy Basin Committee in its study *Bassin de Seine 2050*. This area contains 30 % of France's population and 40 % of its industrial activity. Evaluation of the impact of climate change (more rain in winter, less in summer, and increased temperatures) dismisses the threat of a catastrophe but highlights the problem of more frequent droughts. The models developed have also proved the viability of environmentally friendly agriculture, as well as its positive effects on water quality and surrounding nature, at the cost of a moderate reduction in agricultural yields. And is it a problem if reason arose out of anxiety about the future?

Despite major uncertainties, we need to rethink ways of managing hydrographic basins, and more generally, to revise the principal planning documents in the light of new climate data. As Francisco Cubillo, Scientific Director of the Isabel II Canal, said on 10 October 2007 at a meeting about adapting to climate change: "Thinking of the very long-term future while at the same time taking short and medium term decisions is a sort of revolution. A decision by the Spanish government on water planning has established a compulsory regulation to fix quota reductions in the light of climate change. Hence 5 % less in supply due to climate in the centre of Spain leads to a 10 % reduction in guaranteed services... Every water system has its scheme for distributing its investments, part preventative, part curative, but the climate problem requires a new definition of the proportions and more flexibility."

In 2004 the Escaut[6] Basin communities set up a long-term decision making process, based on integrated management of the area, to make

6. The Escaut is 430 km long; the river rises in the department of Aisne and flows across France, Belgium and Holland before running into the North Sea.

plans for security, economic activities and the environment. The term set is 2030, which is sufficiently far off not to be too emotive, with financial decisions to be taken from 2010 onwards. It is true that the Escaut district is one of the areas of Europe with the weakest renewable water resources, because of its population density and highly developed industrial and agricultural activities.

It is not a question of fighting against nature, but of allying ourselves with nature in order to escape from the climate change trap. Profound climate modification, such as increasing rainfall variability and the multiplication of extreme climate events, but also economic and demographic growth that raise the demand for water, call for stricter management of both resources and risks. It will become essential to create tools for regional planning, together with flexible water management to deal with the risks. It will be necessary to build water infrastructures capable of coping with the broad range of climatic events. It will also be necessary to change pricing policies, and to distinguish in insurance systems between individual provision and national solidarity.

THE WATER WARS WILL NOT HAPPEN

So will the next century bring "water wars"? The press keeps repeating the mantra that in the twenty-first century water will become a resource as coveted as oil was in the twentieth century, and give rise to conflicts that may degenerate into wars. Indeed, the current state of affairs is worrying: climatic change brings water supply changes, including over-exploitation of available resources and increasing demand for water. Industrial pollution sometimes crosses frontiers, as happened in 2005 around the Chinese city of Harbin, on the River Songhua, which then flowed into Russia.

According to the 1997 United Nations Convention on International Water Courses, water is a natural resource, shared to the extent that its use in one country has effects on its use in another. Two thirds of the great world rivers run through several countries. There are 270 river basins that cross frontiers. They increase the economic and ecological interdependence of the countries they cross. The Nile, at 6,671 km our planet's longest river, flows through ten countries. India, China, Nepal, Bangladesh and Bhutan share the waters of the Ganges or Brahmaputra basins. The River Mekong flows through six countries.

The way in which this strategic resource is shared – or not – can give rise to political tensions between countries that share a river. Lack of water aggravates the conflict between the Israelis and Palestinians.

Sharing the water of the Jordan is a major geopolitical issue between Lebanon, Syria, Jordan, Israel and the West Bank. Disagreements between Turkey, Iraq and Syria about dam building and disputes between Egypt and Sudan and Ethiopia on water volumes drawn from the Nile, upset diplomatic relations. Lacking water, the United States puts pressure on Canada to supply them with more. Hostilities arising from scarcity also operate within countries. In Spain, sharing river water between regions with plenty like Aragon, and regions short of water like Andalusia or Catalonia, gives rise to violent political confrontations.

So must we conclude that the ravages of climate change and growing needs will lead to the outbreak of "water wars"? It seems unlikely that these conflicts will lead to actual wars. However, they will exacerbate existing inter-state disputes over water resources and give rise to new ones. And where water is already scarce, if such scarcity is aggravated by withdrawals of the available water for producing bio fuel from agriculture, this will inevitably lead to a hardening of international relations.

Control of fresh water has been the cause of many disputes, but they have usually been resolved without resorting to arms. Despite their successive wars over Kashmir, cooperation between India and Pakistan over the Indus has not been interrupted. Likewise Kazakhstan, Turkmenistan, Kyrgyzstan, Uzbekistan and Tajikistan have set up, with high and low points, a highly original exchange system, involving water for natural gas or coal, to soften disputes over sharing the waters of the Syr-Daria and the Amou-Daria and to keep the peace, despite the loss of the Aral Sea. The truth is that in human history, water management has given rise to cooperation rather than conflicts. Whatever confrontationists and catastrophists maintain, over the last 50 years, there have been 200 inter-state water treaties, as against 37, mostly minor, disputes.

It is very unusual for water to cause people to fight. According to Aaron Wolf, a geographer at the University of Oregon, and director of a project on trans-frontier water conflicts: "The only recorded incident of an outright war over water was 4,500 years ago between two Mesopotamian city-states over the Tigris-Euphrates in the region we now call southern Iraq. Since then, you find water exacerbating relations at the international scale... Strategically, water wars don't make sense. You cannot increase your water resources by going to war with a neighbour unless you are willing to capture the entire watershed, depopulate it and not expect a tremendous retaliation." He adds: "You must distinguish between water as a source of conflict, as a resource or as a weapon of war. We've gone to war over oil. Yet you wouldn't put that event in the same category as the military use of a flame thrower or even napalm[7]." As it indicates in its 2006 Global Human Development Report, the UNDP does not believe that we are looking to a future of generalised wars over water. Neither does Bernard Barraqué, a research director at the CNRS (French National Centre for Scientific Research) and a water specialist, who explains: "Apparently, the king of Jordan's brother has said: 'In the nineteenth century we fought for gold, in the twentieth century we fought for oil. In the twenty-first century we will fight for water.' I don't believe it. The water war will not take place. We may fight with water but not for water... History shows us that people often manage a peaceful sharing of water[8]."

Although agreements on sharing water resources have existed for a long time – for example, the agreement to share the waters of the Aras river between Iran and Russia, which goes back to 1921 – there is very little agreement about fighting pollution, the management of aquifers, and the integrated management of inter-state basins. Anti-pollution conventions, following the example of the one signed in Berne in 1976 by all the Rhine states, are quite rare.

Whether it be to share the resource or manage natural catastrophes, such as droughts and floods, water forces states to improve their cooperation. The idea of "water-solidarity" has been current for the last few years in international vocabulary. It will be used more and more. Indeed, if the impacts of the climate change that is taking place are not collectively discussed well before their effects are felt, there is a high risk of seeing increased friction in trans-frontier basins or mounting accusations of unfair resource-sharing. But consultation will not foresee everything or sort out everything. We must also develop attitudes leading to a more reasonable use of water. Economical management by each nation is vital. Reducing water withdrawals, decreasing sewage dumping in the natural environment, valuing all the available resources are all integral to this "water solidarity" and to an adaptation to climate change.

A FUTURE WITH MORE CLIMATIC DISASTERS

Over the last few years the number, scale and importance of natural disasters linked to water have resulted in increased loss of human lives and have brutally lowered the living conditions of the populations concerned. The main victims are poor people in poor countries. Climate

----

7. "The Water War will not Take Place", conversation with Aaron Wolf by Amy Otchet, *UNESCO Courier,* October 2001.
8. Bernard Barraqué, "L'eau doit rester une ressource partagée", *La Recherche,* no. 421, July-August 2008.

change will make storms, cyclones and floods more frequent. Reducing the impact of these natural disasters has become an international task, particularly since it was included in the UN Secretary General's Advisory Board on Water and Sanitation's Hashimoto action plan. The plan has a twofold aim: to reduce loss of human lives in disasters by prevention policies, and to organise water and sanitation provision during and after a disaster. In 2008, Loïc Fauchon, president of the World Water Council, appealed for a reduction by half in the number of disaster victims of climate disasters such as cyclones.

"Climate change is a threat which can bring us together if we are wise enough not to let it drive us apart" said Margaret Beckett, the British Foreign Secretary. Everyone involved and each water-user must think of what his or her own contribution can be towards adaptation and mitigation policies, in particular by decreasing energy consumption. These policies are more "local" than "global". However, each hydrographic basin must be prepared. Up to a point and although the problems often have a different scope, the same goes for water as for energy. In order to reduce greenhouse gases, there is no need to wait until the USA ratifies the Kyoto protocol: each inhabitant of the planet, each local authority, each industry, each country, can take practical steps as of now, and many have already begun to do so. The same goes for water: each citizen, each municipality, each group of communities in the same hydrographic basin, possess, at their own level, levers that they can decide here and now to pull to prepare for water changes, without waiting for their country's government to proclaim a general mobilisation.

Climate change makes the water question even more crucial than before, but – and this is a puzzling paradox – at the same time it eclipses it. For if global warming is an obvious priority for our societies, does that mean we have to make this question the only focus for debates, as some people seem to want? Over the past few years, by attracting a large measure of the indignation potential, concerns about climate have relegated other subjects to the background, especially sewage treatment and access to water for the majority. One subject should not completely take over from the other. With climate change, the international community gives the impression that it has found a new subject of concern, at the risk of riding roughshod over too many other commitments they have undertaken but not kept. So must we conclude that the agenda for international priorities is now dominated by the media? Why, on the one hand, demand management of the uncertain effects of climate change, and, on the other, refuse to mobilise energies for practical and ancient tragedies, such as lack of drinking water, that kill people every day? On this subject, Loïc Fauchon, President of the World

Water Council, stresses that "numerous problems, population growth, development aid, water pollution pre-exist climate change, which should not become a scapegoat allowing us to forget our errors and omissions." That is why the policy debates of the 5th World Water Forum which will take place in Istanbul in 2009, will concern themselves with these global changes affecting water management that cannot be put down to the impact of climate change alone. There is no shame in prioritising the solution of known problems gravely affecting the present generation, rather than trying to deal, first and foremost, with distant uncertainties for future generations.

II. TWO MISPLACED "GOOD IDEAS": "FREE WATER" AND "USER PAYS FOR ALL"

FREE WATER: A REVIEW OF RECENT HISTORY

Free water in general is one of the mistaken "good ideas" that keeps coming up in debates. It has reappeared recently in debates about water access for all, particularly in the context of the millennium commitments. Ultimately, however, large-scale application of the concept of free water involves an enormous waste of resources, because it takes away responsibility from the consumer. Furthermore, it deprives operators of the finances necessary to maintain a quality service. Free water for all is an economically and environmentally unsustainable strategy. In the long term, residents have free water but there is nothing left!

Let us look back over recent history. In Prague, under the communist regime, the water service was delivered at a very low – flat-rate – price. The message given to consumers was that water was available in apparently unlimited quantities. For lack of sufficient revenue, the service quality declined. After the political regime change, the Prague water service had to make important investments to repair infrastructures and it revised its tariff policy in order to defray the costs of the service more fully. Between 1990 and 2007 the price of water increased 18 times in real terms. The change from practically free water to water charged at an average price of €1.9 per m³ brought residents' consumption back in line with normal levels for Europe. During that period the total volumes consumed were halved, from 154.4 million m³ in 1997 to 87.4 million m³ in 2007.

In Romania, in Ploesti, a town of 233,000 inhabitants, the introduction of meters and the abolition of fixed-rate bills, alerted people to water leakage in the internal plumbing of dwellings from such things as faulty washers. It also led to a profound change in habits: when water was nearly free, it was used to cool drinks, defrost meat, and so on. The introduction of meters and the change to pricing by volume led to a sharp decrease in daily consumption. In just four years, from 1999 to 2003, consumption dropped from 350 litres per person to 135 litres. Following similar policies, most Romanian towns showed the same result, including Bucharest with 1.8 million inhabitants.

Although different, the case of India is instructive. Today in the second most highly populated country in the world, none of the large cities is able to provide its inhabitants with water for 24 hours a day, 7 days a week. The main reason for this is urban growth. But the decision to charge for water at a very low price, aggravated by the irregularity and small scale of public subsidies designed to compensate for the costs not covered by tariffs, has greatly increased the difficulties encountered by the water services. According to the researchers Raghupthi and Foster, in 2002, most Indian towns charged for water at a price equivalent to only 10 % of the costs of operation and maintenance. In India, as has been said of other South Asian countries, "people have a willingness to pay but governments don't have a willingness to charge." One of the responses of Indian towns was to reduce supply times. At first, this was programmed for a few hours a day, but then periods with the water cut off were lengthened. Intermittent supply, provision for two or three hours a day at most, district by district, for the better off, has often become the norm for users. Many now organise their own storage, while plumbing engineers and architects make provision for underground storage and water tanks on each roof. However, this strategy tacitly thrusts the costs back onto individuals who pay for individual water tanks and pumping and treatment systems for each building. In the end, this additional cost makes water much more expensive than if it was delivered 24 hours a day by an efficiently managed public system.

So, apart from situations of extreme poverty, such as in refugee camps, it is desirable to make people pay for their water service. That does not mean that everyone should pay the same price: in order to alleviate the cost of water supply to people on low incomes, social tariffs, as in most African countries, or direct aid systems (on the Chilean model) can be introduced.

THE MYTH OF A FREE WATER SERVICE

"God gave us water but he did not supply the pipes." This witticism has the merit of reminding us of a truth that cannot be ignored: although water – humanity's common good – has no price, the service to provide it does. The price has to cover the costs of water treatment, management and maintenance of the network, infrastructure investment, and interest on debt. Once we realise that a public service costs money, then it is not and cannot be free. If the local authority decides to provide it free to certain people, it simply means that its cost is borne by others, so that it appears that the service is free for those who do not pay. Just as a water service has a cost, so does the right to water. Therefore this cost has to be borne by someone, if consumers of the service cannot pay it in full.

In this context it sounds strange to hear that water is free. If they are talking about centuries of free water, particularly in rural areas, that is not the same thing at all. In one case it was a question of water, not made drinkable or purified, that people went to draw through their own efforts. In the other it is a question of water made drinkable (and so with no risk to health), available 24 hours a day and delivered to the home, and then purified after use.

If free provision of water means funding the whole system by taxation, it ignores the painful experiences of the past. It goes back to much criticised practices in towns of the Soviet Union, which are still operating in some agricultural areas of the world. At a time when, in many regions, we need to confront water scarcity, it sounds like an invitation to waste.

If "free water" ideas are limited to the application of preferential tariffs to poorer populations, that already happens in very many countries. Town mayors are not in the habit of liquidating the social dimension of their water service. Nevertheless, it is true that many tariffs do not correspond well to the classic aims of water pricing. The task is then to improve tariff policies and optimise various solutions, including tariff solutions and others, to decrease the cost of water provision and respond better to basic demands.

But then if free water is provided to satisfy basic needs, we see a paradox: this free water is demanded in countries whose inhabitants are able to pay for water, and where social systems, however imperfect, are relatively effective. So the question arises: why make only water free? Why focus on something that represents 0.8 % of household expenditure (in France) and not on housing, which represents 15-25 % of the budget? Those who reject payment for water by the customer and declare that water should be free, refuse to water provision what they willingly grant to other goods and services that are also indispensable, such as food, clothing and housing.

If we turn to developing countries, whether the water supply is free or not is not the point. Those who do not have access to public networks, and therefore pay for their water ten or twenty times more dearly than those who are connected, do not ask for water to be free: they want to be connected, that is, to receive the same service as other citizens. They are even prepared to pay more for water, if that means a real improvement in the service. The current underpricing for water and sanitation services only marginally benefits the poor. The same would go for free water delivered through the public network. Maurice Bernard, from the Water and Sanitation Division of the French Development Agency, has a clear opinion on the subject: "Supplying water free creates exclusions: it benefits those who already receive a public water service to the detriment of those who do not[1]."

As the geographer Yves Lacoste notes: "Certain political movements, particularly Greens, demand that water should be considered an indispensable and inalienable right, that it should be freely distributed in sufficient quantity to all the inhabitants of poor countries... That is an attractive suggestion, but opposing the financial approach to the problem of water distribution holds back the implementation of large hydraulic programmes, which require the investment of large amounts of capital[2]."

The case of South Africa is worth examining. In 2000 the government brought in a policy of free water for the first 6 m^3 per household per month. At the same time it raised the tariff for the middle consumption block of 6-30 m^3 per month and for the higher consumption block (above 30 m^3). Certainly, this system reduces the cost of water provision for poor families, but has several drawbacks. The free provision of the first 6 m^3 of water per month benefits all, including the middle class and the rich, who do not need it. In general, the water services have had to raise prices several times for the middle and higher classes, in order to maintain their receipts. They have also seen a progressive decrease in the average rate of bill payment since the introduction of the free water in the first tariff block.

THE OPPOSITE OF FREE SUPPLY: THE USER PAYS FOR ALL

Recovering the full cost of the water service from the consumer is the diametrically opposite error to recovering no costs through the water bill, and in developing countries, full cost recovery through water tariffs is another a misplaced "good idea". The principle "user pays for water", that operates in some developed countries, is unrealistic in developing countries. The investments required are much too great to be borne solely by consumers of the service. In developing countries, the idea of "acceptable cost recovery" needs to replace the principle of "full cost recovery".

----

1. Maurice Bernard, "Aide publique au développement et objectifs du Millénaire", *Revue Politique et Parlementaire*, no.1043, L'eau: la guerre aura-t-elle lieu?, April-June 2007.
2. Yves Lacoste, *L'eau dans le monde. Les batailles pour la vie*, Éditions Larousse, Petite encyclopédie collection, 2008.

In passing, it is worth noting that even in developed countries, full cost recovery has some important exceptions. In Holland, part of the purification costs are financed from taxes and not paid for by water and sanitation service customers. In Rome and Athens, the construction of some drinking water and sanitation plants were subsidised by money from taxation: so these investment costs are not covered by the price of water. In France the principle of cost recovery from the customer is applied rigorously, with a legal dispensation for local authorities with less than 3,000 inhabitants. But it has not always been so. At the beginning of the twentieth century the user only paid half the cost. The taxpayer and the following generations paid for the other half. Thus, drinking water networks were financed over three generations, then sanitation systems over the next three generations. In all, the construction of infrastructures was financed over six generations. The problem for developing countries is the same as the one that faced France and other European countries 150 years ago. So developed countries are not in any position to advise developing countries to adopt, from one day to the next, a method for financing infrastructures that they themselves took several generations to apply, nor to require them to accomplish in just a decade the revolution in water supply that took Europe more than a century.

Of course, it is understandable that the principle of full cost recovery from the customer is attractive to service managers, both public and private. It offers the security of financial autonomy for the service and greater transparency with regard to its price. Financial autonomy is gained, because the water service is no longer at the mercy of public subsidy, which might be withdrawn or cut for extraneous political reasons. Financial transparency is gained because when the user pays for all the water – and pays for nothing but water – average prices reflect average costs, and this tells the consumer the real cost of the service. Although this principle cannot be applied to all towns today, it is desirable in the long term to internalise all the costs that can be borne by users. Once the main infrastructures have been built, and the phase of intensive equipment construction has passed into the phase of regular maintenance and renewal, once populations are stabilised, then the gradual establishment of cost recovery can be envisaged. But this will always be a choice for the public authorities.

THE HAPPY MEDIUM: A SOCIABLY AFFORDABLE
WATER PRICE

One of the basic principles for tariff setting is not free water, but water affordability. Access to an essential service requires an affordable price. According to public authorities or international organisations (UNDP, WHO, The World Bank, etc.) households should not spend more than 3 % to 5 % of their income on a water and sanitation service. For example, in Indonesia, one of the principles of tariff policy for the public water services requires domestic water bills not to exceed 4 % of basic family income.

What really happens? In Gabon, households connected to the water network spend, on average, 0.55 % of their income to get water from the public network, as compared with an average of 1.4 % in other sub-Saharan African countries (excluding South Africa)[3]. But in many African towns, people who are not connected to the network devote a much larger proportion of their income to obtaining water, sometimes more than 10 %[4]. The same goes for Nicaragua and Jamaica. And this percentage does not take account of the supplementary "costs", in terms of water-borne diseases and time lost in getting water. As poor people put it themselves: "It costs a lot to be poor."

On the other hand, in European capitals, water bills as a proportion of household income vary between 0.3 % and 0.8 %, and sanitation between 0.3 % and 1 %. The proportion is highest in Amsterdam, reaching 1.8 % of household incomes for both services[5].

How can the price of water be kept affordable when its investment needs would take it above the threshold of 3-5 % of household incomes? Firstly, finance must come from users being made to pay what they can afford, and secondly, by organising a transfer of funds from outside the water service. This might be from local or national tax revenues, revenues from the electricity service (as happens in Senegal and Madagascar), revenues deriving if necessary from other municipal services (as in the German *Stadtwerke*[6]), or international development aid.

----

3. According to a study carried out by the International Consulting Economists' Association (ICEA) in 2007.
4. Bertrand Dardenne, *The role of the private sector in peri-urban or rural water and sanitation services in emerging countries,* working document, OECD World Sustainable Development Forum, Paris, November 2006.
5. *Analysis of water and sanitation services in eight European capitals with regard to sustainable development,* BIPE (Bureau d'Informations et de Prévisions Economiques) Study, 2006.
6. The *Stadtwerke* are German community enterprises funded by public or mixed capital. They are empowered to manage several activities simultaneously, such as water, energy, transport and refuse collection.

12. Water and sanitation in household budgets

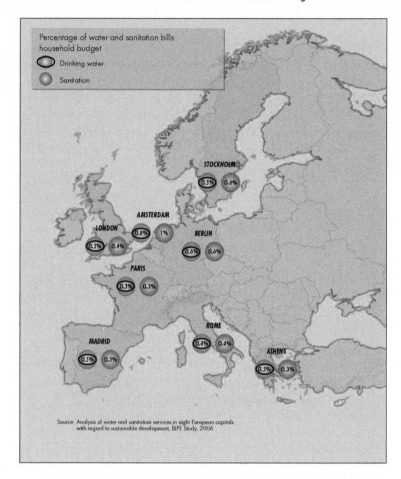

Percentage of water and sanitation bills
household budget

◯ Drinking water

● Sanitation

STOCKHOLM 0.5% 0.6%

AMSTERDAM 0.8% 1%

LONDON 0.5% 0.4%

BERLIN 0.6% 0.6%

PARIS 0.3% 0.3%

ROME 0.4% 0.4%

MADRID 0.5% 0.3%

ATHENS 0.5% 0.3%

Source: Analysis of water and sanitation services in eight European capitals
with regard to sustainable development, BIPE Study, 2006

III. THE PRIVATE SECTOR: TOO MUCH OR TOO LITTLE INVOLVEMENT?

Whenever it plays a part in the management of water services, the private sector, abhorred by some whatever it does or doesn't do, for others the panacea, gives sometimes rise to violent controversies. In France, civil society is riddled with many debates about the ways to organise the public water service and the role of private operators. Nevertheless, 83 % of French people are happy with their water service, according to a survey carried out by 2008 SOFRES-C.I. Eau. France is the cradle of delegated water management, which continues to attract others abroad. Would the pragmatic and demanding Chinese leaders turn to delegated water management if it did not suit their fellow citizens? Would the Czech elected representatives have resorted so often to the private sector if their communities had not derived tangible advantages from it?

On a world scale, private businesses manage less than 10 % of water services and less than 3 % in developing countries. That is the reality. Is that too much or too little? In fact, this question does not make a lot of sense. Private enterprise knows that it is up for questioning and will be so for a long time. The particular questions asked of it are about its role in relation to the public authorities, about the prices of the services it provides, its interventions at international level and in developing countries.

THE MANY FORMS OF COOPERATION BETWEEN PUBLIC AND PRIVATE

The common will of local authorities and professional operators to improve essential services has been expressed in long-lasting partnerships, based on a relationship of trust. This form of collaboration between a local authority or state, on the one hand, and a private company, on the other, brings together a public service mission and the efficiency of a private operator. It enables the public authorities to benefit from the expertise brought by a professional operator, while at the same time keeping control of strategic aims and water policy. In the water sector, the first collaborations of this kind took place in France in the middle of the nineteenth century. So this practice has a proven 150 years of experience.

Such partnerships place the local authority at the heart of the decision making and the professional at the centre of the mission. The sharing of roles between public authority and water service manager is conducted with great rigour. At a time when the need is felt to develop new modes of governance, the flexibility of public-private partnerships, which have spread all over the world, while adapting to the specific expectations of each community, is an asset. In practice, these partnerships take very varied forms: management contracts, leasing, concessions, contracts to build and operate an infrastructure (for example, BOT: Build Operate Transfer), etc.

In the case of delegated management, the local authority entrusts a private operator with the provision of water services after putting it out to tender and then offering a fixed term contract. Prices are fixed for the whole duration of the contract; the operator has to take on the operational risks, and respect targets. An important point is that private operators only become involved when requested to do so by a public authority.

Those who experience and operate under a service delegation contract know that the dialogue between the local authority and the operator is constant and vital, whether it be on technical and economic matters or such diverse matters as the programming of works, water quality or customer management. Beyond the operator's professionalism, successful service delegation depends both on a contract clarifying responsibilities, and on trust between the players that has to be maintained on a daily basis.

In each country, drinking water and sanitation services are provided by a specific organisation that arises out of the country's particular history and geographical, social and institutional conditions. However, there are three major parameters by which to compare them: ownership of infrastructures, sharing of risks and responsibilities, and price setting.

Let us begin with the infrastructures. In all the countries of the world, with the exception of England and Wales, Chile and about 10 % of services in the USA, infrastructures belong to public bodies, even when a private operator is involved. In practice, they are owned, directly or indirectly, by the public authority, sometimes through the intermediary of a publicly funded company.

The sharing of responsibilities leads to a threefold division. Firstly there is set-up and structural activities, such as the organisation of the service, controlling it and the definition of tariff policy. Secondly, there are the activities concerned with general management of the service, for example, customer management or infrastructure policy. Thirdly, there are delivery tasks, such as meter reading, repairing of leaks, and so on. If the public authority in charge of organising water and sanitation services wants to fulfil all these roles, the service is then managed by the public sector. If not, the authority can decide to use the services of one of more private companies. According to the degree of autonomy that the public authority wants to grant the private sector, it will entrust only delivery tasks to the private company or it may further ask the company to take on management tasks. Then there will be delegated management with leasing contracts (when the operator takes over the daily provision of the service) or concession contracts (when as well as this, the operator must construct and finance new works which, even in this case, will remain the property of the public authority). The autonomy granted to the private operator in no way means a lack of regulation. In all these cases, the public authority defines the strategy to be put into practice and the goals to be reached, it retains the power to control and supervise the service, and it decides on the tariffs to be applied. In sum, the private partner just implements the water policy defined by the public authority.

Price setting can be by to two different mechanisms. Either it is determined in advance: in particular, this is the case with delegated management, which gives an incentive to control costs. Or prices are set *a posteriori* as a result of the costs incurred, as mostly happens with public sector management. But either way, the public authority, never the private operator, decides the price. In France, Germany and Spain, it is up to the competent local authority to set the price of water (under whatever form of management), whereas in England, the national regulator, OFWAT, does this task.

PRIVATISATION: SO FREQUENTLY DEBATED, SO RARELY PRACTICED

In order to realise that service delegation is not privatisation, we need only go to England and Wales. During the 1980s, reform of water management, orchestrated by Margaret Thatcher's government, led to complete privatisation of the water service. Since then, the English system has relied on the transfer of ownership of public infrastructures to private companies, who must run them in a way to ensure the provision of a water service, and the establishment of a public regulator charged with overseeing the proper functioning of the service. The regulator has set up very precise assessment systems to measure the performance of the water services and ensure the system's transparency. However, one thing is remarkable: the regulator is not elected. The regulator is a public authority independent of political power. Why make

13. The sharing of functions between public and private

Power over set-up and structural activities	- Initiate the creation of the service - Assess needs to be met and define principles and overall budgetary requirements - Decide on a water policy and the principal developments of the service - Choose between direct and delegated management - Exercise control over the management of the service - Fix the price of water - Actions requiring some exercise of public compulsion (in different states this might include policing of networks, expropriation, rooting of pipelines...)	**Public body**
Power of management and exploitation	- Define ways of supplying the service, the means, organizations to be set up, including human resources, technical equipment, etc. - Build and finance new works - Take initiatives and make proposals for improving the service - Manage research, updating and training	
Power over delivery tasks	- Train and pay staff - Maintain machines and equipment - Ensure repairs and maintenance are carried out - Carry out administrative tasks - Manage relations with customers and suppliers - Monitor water quality	**Private partner**

such a choice? Doubtless, this relates to a different conception of regulation, in which notions of "public" and "democratic" are more sharply distinguished. According to this system's promoters, the important thing was to disconnect water management from the political electoral cycle with the temptations that might ensue from it. On this view, an unelected public regulator is better able to promote the long-term collective interest.

The differences are stark between the English approach, which has been copied by only one country, Chile, and the French approach, which very many countries have followed. In England there is centralised regulation at national level, in France there is decentralised regulation by local authorities. English elected representatives are deprived of any role in the regulation of the water service, whereas their French equivalents have a central one. North of the Channel, public authorities are dispossessed of their assets, whereas they retain full ownership of them in France. In England and Wales the price of water is fixed by a distant national regulator, whereas in France elected local representatives fix the price of water.

Whatever the preferences for the one or the other, the English system gives rise to two fundamental problems. The first is the difficulty of reversing the method of water management, since the public authority would have to buy back the infrastructures from the private owner. The second difficulty is the lack of a way for local political representatives to respond rapidly to bad decisions that may be taken by private operators. There is a third problem, which makes this type of private sector involvement unsuitable for developing countries: apart from removing ownership of essential infrastructures from the public authority, privatisation needlessly increases the price of water. Indeed, when a private operator buys public infrastructures, he recovers their cost in the average water price, so as time goes by he will recoup his investment.

Delegation of a public service is not privatisation, because it is not accompanied by the transfer of ownership of plants and public networks to private enterprise. In more than 99 % of cases, when the term "privatisation" is applied to water or sanitation services, it is used wrongly. Throughout the world, the privatisation of water services exists only in three countries and they remain exceptions. The UNDP also draws attention to the fact that the term privatisation is not appropriate for all public-private partnerships: "the diversity in public-private partnerships cautions against lumping all private sector involvement under the general heading of 'privatization'[1]."

TO DELEGATE OR NOT TO DELEGATE

France has been called the country of delegated management: 52 % of communities, representing 72 % of the population, receive their drinking water from a private operator, and 38 % of communities, representing 55 % of the population, their sanitation. Direct public management is the basic rule in France, as it is a choice that requires no deliberation or justification by elected representatives to their fellow citizens, unlike delegated management, which requires an explicit decision by the elected representatives. Nevertheless, delegated management is in France the most frequent system used.

France is first and foremost the country of freedom of choice about management: public management or management delegated to a private company. Sometimes these two management methods co-exist within the same community, with for example, sanitation operated by public sector and drinking water supplied by the private sector. Throughout the world this freedom of choice between management methods is far from being general. In many countries the public authorities have no free choice on the subject. Restriction or complete lack of

----
1. 2006 *Global Human Development Report*, UNDP, *op. cit.*

choice can happen in two ways: the imposition of a private operator, together with its mode of operation, as in England (with initial competition when the choice of operator is made); or a legal or constitutional ban on the provision of drinking water by private operators. The latter solution, which already exists in Uruguay, and is on the way in Holland, binds local populations to their public operator, whether it is good or bad.

Comparison between services responds to the expectations of the population. This need for comparison is all the more legitimate, since focusing on prices alone can conceal actual service levels. Regions have very different conditions, according to their geography and human activities, which determine water provision and pollution management. Driven by recurrent arguments about water prices, French legislation has established a programme of comparison based on objective data. The National Office for Water and the Aquatic Environment (ONEMA), set up by the water and aquatic environment law of 2006, has been given the task of drawing up, between now and the end of 2008, an instrument for monitoring performance. This will enable communities to evaluate their services better, and to compare them with other communities where similar conditions apply.

When we ask elected representatives for the reasons that led them to opt for delegated management, they often cite as their first reason the transfer of technical and legal responsibility to a professional operator. A study based on France, and carried out by the Boston Consulting Group (BCG) in 2006, throws light on the debate: delegations are much more frequent where technical operating conditions are complex. Where the resource is derived from surface water, which is more polluted and more difficult to treat, the local authority will more often call in a private operator. Thus, for the delegated water services, the quantity of good quality source water available is half that for services under public management. A further indicator shows that delegated services manage proportionally 2.5 times more coastal communities, where pollution concentrates and there are seasonal peaks in consumption and very high quality standards for bathing water. In short, the more complicated the situation, the more often it is delegated.

A further reason for the decision to go for delegated management lies in human resources management. Men and women are at the heart of the service tasks. A large private business has more room for manoeuvre than a public operator; it can establish more dynamic human resource management, give workers more responsibility, offer a variety of professional development, and develop more opportunities for training and passing on know-how. In 1992 the city of Caen delegated the management of its water and sanitation services to the company Générale des Eaux, which later became Veolia Water. In 1997, each officer had to agree to a new five year secondment or a return to the town hall. All of them chose to remain within the service. When they were asked why, they replied: "more interesting work"; "greater autonomy given to managerial staff"; "each person can develop according to his aspirations and skills"; "access to a national and international skills network".

The performance of a service is assessed using many parameters: rate of compliance with standards, network yield, customer service management, reliability of the service, human resource management, cost of the service, investments envisaged, etc. In order to prepare for their decision, most local authorities begin by comparing the advantages and disadvantages of the two options, any results obtained in the past and the goals to be reached. Jean-Michel Héry, deputy mayor of Rennes, stressed this in a conference organised in 2006 by the consumers' association, Indecosa-CGT[2]. He said: "At the end of a contract we commissioned a researcher to compare direct public management with delegation, so as to clarify the choices put before our elected representatives. The financial advantage favoured direct public management by 2 %. But what concerned us was the quality of water in the tap. This was not a foregone conclusion in Rennes, where we have had problems with water quality. The decisive factors in determining the option for delegation were the global aspect of the problem, and the priority given to expertise and efficiency of service."

In the end, the reasons leading to delegated management may be numerous. They may be technical, as when it is a question of benefiting from the expertise and knowledge of an experienced professional, or gaining legal security in the sensitive areas of public health and protection of the environment, or becoming part of an international network, so as to benefit from greater expertise or innovation. Reasons may also have to do with human resource management and management in general, as we have just seen, but may also be related to the need to control the costs of the service, optimise purchases, and have long-term transparency in the pricing of water. The choice of delegated management may also be linked to the achievement and financing of new investments, thus smoothing the impact on prices, and optimising the cost/performance ratio.

----
2. "Un droit, l'eau c'est la vie", Conference of Indecosa-CGT, Bobigny, 17 octobre 2006.

PUBLIC-PRIVATE PARTNERSHIP: FROM FASHIONABLE CRAZE TO MATURITY

The United Nations declared the 1980s the "Water and Sanitation Decade". but the results did not come up to expectations. There were few real advances, but at the beginning of the 1990s, the need was felt to reconsider the question of water. In 1992, the Dublin Conference on Water and the Environment marked a breakthrough. All were agreed in recognising that local public authorities in emerging and developing countries could not deal on their own with the challenge of providing water and sanitation. The sorry state of certain services required a thorough reform of water "governance". The key idea promoted at Dublin was to introduce more economic rationality into water management, recognising two indispensable imperatives: democratic control and efficient management.

So public-private partnerships became one of the solutions chosen. In fact, of course, the private sector brings expertise, and the expertise of such big companies as Veolia, Suez, Saur or RWE (Rheinisch-West-fälisches Elektrizitätswerk) is internationally recognised. Considered more effective in satisfying the needs of consumers, the private sector boasts a reputation for greater efficiency than the public sector. Moreover, private enterprise offers access to diverse sources of finance, either directly by injection of the company's capital, or through the flow of funds released by operating profits; or indirectly, as the private sector is often better able to raise local or foreign capital, whether it comes from banks, financial institutions or insurance companies. By adding its credibility, the private operator brings the local authority greater access to various financial resources, and gives the local community an important lever.

The desire to take into better account the economic value of services linked to water also helped to justify the involvement of the private sector. It starts with a basic assumption: for the service to be well run over the long term, the price of water should not become totally divorced from its real cost. To ensure the sustainability of this essential service, it is vital to introduce an economic rationale. Going to the private sector has attracted the international community because the previous public authority management often ran into the problems of extending the service and financing it, of quality and of balancing the books. Water distribution is an emblematic public service and it is tempting for some politicians responsible for it to campaign on the subject of water price. Promising to decrease it – or refusing to increase it even though that may be necessary to carry out investments to protect the environment – can be a seductive option. But by opting for it, many water services, especially in developing counties, have found themselves financially stymied. As Bernard Collignon, president of Hydroconseil, a research and engineering company specialising in water in developing countries, stresses, "Many public bodies have thus fallen into chronic deficit and become incapable of financing the smallest works, even the most urgent ones, and even less able to extend the service to meet demand. They no longer serve any but a small percentage of households. By blocking the price of water, governments have fallen into the shocking quandary where the taxes of all are used to subsidise an enterprise that only serves 3 % to 10 % of households[3]."

SNAPSHOTS OF THE LATIN AMERICAN EXPERIENCE

• Buenos Aires

Buenos Aires is decidedly the most widely publicised case of withdrawal by a private operator. When it began in 1993, this drinking water and sanitation concession was the biggest in the world, and it announced extremely ambitious goals. In this city of more than 10 million inhabitants, the first investment plan aimed at reaching 3.5 million new customers, 65 % of whom lived in the poor suburbs of the Argentine capital. There was not enough time to achieve the final goals because the contract was cut short; nonetheless the goals set for the period before the Argentine financial crisis in 2001 were reached. The rate of coverage was significantly improved: for drinking water the rate increased from 70 % in 1993 to 87 % in 2001, and for sanitation the rate increased from 58 % to 64 % over the same period[4]. From the beginning of the contract until 2001, investments amounted to $1,718 million (the contract provided for $1,702 million), water pressure increased by 80 %, the rate of compliance with standards increased to 99 % and the rate of customer satisfaction rose above 90 %[5]. When the contract was cancelled by the Argentine government, more than 2 million residents had been connected to the drinking water network and more than 1 million to the sanitation network: a major, unambiguous advance for the population.

The 2001 Argentine economic crisis led the government, in January 2002, to abandon parity between the American dollar and Argentine peso, which had previously been maintained by law, and the Argentine

----

3. Bernard Collignon, "Les grands bidonvilles africains: la prochaine frontière pour les distributeurs d'eau", *Revue Politique et Parlementaire*, no. 1043, L'eau: la guerre aura-t-elle lieu?, April-June 2007.
4. Bertrand Dardenne, *op. cit.*
5. Guy Canavy, *Les multinationales de l'eau et les marchés du Sud*, Suez Environnement, Collection débats et controverses, GRET, June 2007.

peso fell to a third of a dollar. Being paid in pesos, the concession-holder became incapable of repaying the debt it had incurred in order to invest. Expressed in dollars, the company's income for the water and sanitation service was divided by three, whereas its debt remained unchanged. In March 2006, after four years of fruitless negotiation between the state and the private concession-holder, the Kirchner government then in power refused to increase tariffs so as to compensate the private investor as it was required to do by contract. It ended the contract and created a new business, Agua y Saneamiento Argentinos (AySA), with public capital, giving it far fewer responsibilities than those laid upon the former private concession-holder. Large investments were then not financed by the new company but by the state.

It is worth recalling that the withdrawal of capital from Argentina involved all sorts of companies, both public and private, and in all sectors, including telecommunications and energy. EDF (Electricité de France) – at that time a 100 % publicly owned company – disengaged itself in 2005, selling its Argentine subsidiary Edenor, after having suffered losses of $122 million over the previous three years, and having accumulated a debt of $512 million[6]. Clearly, these Argentine misadventures were neither specific to water nor to private operators. The story of the Buenos Aires contract highlights a major problem. Over a period of 20 to 30 years (the usual length of concessions), many developing countries risk suffering, like Argentina, a major economic and financial crisis. This means it is vital to create contracts – and especially means of finance – that are sufficiently robust to survive such turbulence.

• San Miguel de Tucuman

The region of Tucuman in northwest Argentina is another example. With the concession of water and sanitation services to the private sector, the authorities of the province of Tucuman chose to pass the whole cost of the service on to the users. Shortly after the concession began, and following local elections, the provincial government of Tucuman changed its position, and refused to respect its commitments on the price of water, even though the concession contract provided for large investments. This left the operator, Compañia de Aguas de Aconquija (CAA), a former subsidiary of Veolia[7], with no choice but to demand the termination of the contract by virtue of breach by the contractor. That was in 1997, only two years after it was drawn up, and well

----

6. *La Tribune.fr,* 28 April 2005.
7. To be precise, because of an existing dispute before the split between Vivendi and Veolia, this subsidiary has remained the responsibility of Vivendi Universal.

WATER

WATER'S FALSE FRIENDS

before the devaluation of the Argentine peso and the country's economic collapse. The private company initiated proceedings before the ICSID, the International Centre for Settlement of Investment Disputes. These proceedings were against the Argentine State and they resulted in a halt in the functioning of the concession enforced by the provincial government of Tucuman, and manoeuvres by the province to undermine the project's economic viability. The proceedings were founded on a claim over a violation of the measures agreed in the bilateral Treaty for the protection of investments signed by France and Argentina.

In 2007, the concession-holder was vindicated and the Argentine State was ordered to pay $105 million in damages. In its judgment, the arbitration tribunal considered that the behaviour of Tucuman province, and thereby, of the Argentine State, was in complete breach of the principle of just and equitable treatment, established by the bilateral Franco-Argentine treaty. "On the facts before us, it is only possible to conclude that the Bussi government, improperly and without justification, mounted an illegitimate 'campaign' against the concession, the Concession Agreement, and the 'foreign' concessionaire." The Tribunal also considered that the various actions taken by the province had a devastating effect on the financial results for the concession, naturally driving the concession-holder to terminate the contract. As for the concession-holder, it had been confronted with a case of "international delinquency".

The Tucuman concession raises several principles in relations between public and private partners. First of all, when a public authority puts out the management of a service, it must play the game and allow the private operator to accomplish his mission. But, whatever precautions may be taken by a private company, it is impossible to know with full certainty in advance how his future public partner will behave, since a change in the political majority may lead the public authority to change its mind radically. Secondly, in all delegated management contracts, the delegating authority can, if it wishes, return the service to direct public management. If it decides to do so, it must compensate the concession-holder fairly for the investments made.

• La Paz and El Alto

The La Paz and El Alto concession in Bolivia began in 1997. It was supposed to last thirty years but came to an end after about ten. The company involved, Aguas del Illimani, a subsidiary of Suez, had undertaken to extend the service. El Alto is a town situated over 4,000 m above sea-level, with a more than 5 % rate of demographic growth and about 650,000 inhabitants. Two thirds live below the poverty level. One of the main difficulties for Aguas del Illimani was that a large part of the El

Alto outer-urban area lay on the geographical frontiers of the concession, but were not included in the districts to be served, so the concession-holder had no obligations towards that area. As Bernard Dardenne says in his 2006 study of the role of the private sector in developing countries, "The political opposition against 'water privatisation' focused on the fact that the concession left tens of thousands of poor families out of the process. Thus, the concessionaire looked for a way to mitigate such a disastrous image. The strategy has been based on a partnership between local authorities, community organizations, the private operator and financing institutions. [...] Even with a consistent track record, the concession of La Paz – El Alto remains fragile when the political environment becomes adverse, due to other unsuccessful experiences that occurred in the country (Cochabamba) or due to radical ideological changes in the government[8]." This study stresses that one of the main difficulties came from inhabitants who demanded that their districts should be included within the concession area: they also wanted to benefit from improvements in the service. In other words, they were demanding to be included within the public-private partnership instead of being excluded from it! They wanted more public-private partnership, not less!

Relations between Aguas del Illimani, the public authorities and the population deteriorated in an increasingly tense social context, marked by budgetary austerity, social volatility at the national level, and political instability (the resignation of President Sanchez de Lozada). In January 2007, one year after the election of President Evo Morales, an agreement was reached between the two parties, after a long negotiation. The government and the company decided that the Bolivian State should buy back the concession from the private concession-holder.

Among the various lessons to be learned from the La Paz-El Alto concession, one stands out: a project that does not envisage, at least in the long term, bringing the same improvements to all the people living in the same area, is destined to create huge frustrations capable of thwarting it altogether.

Has the situation improved since the private operator has gone? To give the whole picture, it is instructive to study the development of water services after the withdrawal of the private operator. Bernard Collignon, of Hydroconseil, analyses the situation as follows: "Over the last few years, when an international private operator has withdrawn, several countries have had rapidly to refloat a public enterprise for water distribution (Conakry, Manila, Cochabamba, Bamako, Buenos Aires), which constitutes *de facto* nationalisation. Unfortunately, however, this has not

----

8. Bertrand Dardenne, *op. cit.*

proven to be a reason for rejoicing for the poor. The same causes bring about the same effects. These operators quickly fell into the same pitfalls as before, and the water services at Conakry and Bamako, for instance, have begun to deteriorate[9]."

ARE PUBLIC-PRIVATE PARTNERSHIPS APPROPRIATE FOR DEVELOPING COUNTRIES?

During the 1990s a number of large-scale international water businesses intervened in developing countries, encouraged by financial backers, particularly the World Bank, tired of financing public enterprises, only to lose money and see water services collapse. Between 1997 and 2000, these big private companies invested nearly €25 billion to modernise services and extend networks[10]. However, there was great disappointment over several of these large contracts. Public authorities accused the private companies of not respecting their investment objectives. For their part, the private companies accused the authorities of not having organised conditions for them to do their job properly or ensured the establishment of the anticipated tariffs. So public-private partnerships began to be distrusted. The withdrawal of international corporations became widespread, and became a pretext for some to proclaim the systematic futility, even harmfulness, of intervention by private operators in developing countries.

It is true that improving a water service and extending provision is a real challenge in a developing country. The operator has to do the best he can in a difficult set-up. In rapidly changing towns, management of the water service has to face both social and political difficulties. Even more, there are inevitable technical difficulties involved in modernisation: intrinsic problems in the outer urban areas and shanty towns, which hold up the extension of networks; the poor level of staff training; dilapidation of infrastructures; extremely limited financial resources to improve the service; the lack of a reliable address system, which limits the use of the post for sending out bills. The difficulties are many...

So are public-private partnerships unsuited to developing countries? Not at all. This way of cooperating between the public and private sector can certainly be adapted to developing countries, but it cannot be any kind of partnership in any kind of situation. It would be a pity if public authorities in developing countries were to deprive themselves of this instrument that can be so useful when the conditions for resorting to it are right.

----

9. PPI Database (World Bank – Public-Private Infrastructure Advisory Facilities, PPIAF) cited by Bernard Collignon, *op. cit.*
10. PPI Database (World Bank – PPIAF).

• Learning from mistakes

Although many public-private partnerships have been successful, others have failed. It has to be recognised that on certain contracts the private operators have made mistakes. These can include an over ambitious investment plan in relation to what the financial or social context permitted or an over ambitious timetable for extending networks. The company may seek to apply literally the principle of full cost recovery from users (following the recommendations of international financial institutions). There may have been excessive optimism about legal safeguards in certain countries, or about expecting the delegating authorities to behave as genuine partners in case of difficulty. The company might have experienced great exposure to the risks of exchange rate turbulence, with a debt contracted in foreign currency and payments made in local money; or it may have underestimated the importance of acceptance by the local inhabitants, although this is a vital social factor for the success of a public-private partnership, etc.

Private operators have learned a number of lessons, both from successful and from failed interventions in developing countries. Old lessons have also been relearned from recent experiences. We have, however, learned that nothing lasting can be constructed without the assent of the population. In one way or another, they must be consulted about the methods chosen to manage their water services. We now know that concessions cannot work unless there is a mature local financial market, capable of accepting sub-sovereign risks[11]. Private operators (that is, those who did not know already) and financial organisations who recommend them have realised that the principle of full recovery of costs from the user is beyond the capability of most developing countries.

We also know that "everything is not possible all at once". When a town grows too quickly, or when a service has been neglected for too long, everything cannot be put right immediately. Investment programmes should rely more on the real state of the infrastructures than on the authorities' understandable desire to extend the service as quickly as possible. That has a consequence. While starting to build infrastructures or while working to improve a dilapidated system, the public authorities and the private operators must both make an effort to "manage impatience". For example they can resort to interim solutions, such as standpipes, to provide an improved water service, before the network extension to the outer-urban areas makes it possible to connect each household individually. We have also learned that too many invitations to tender are made based on incorrect figures – such

---

11. That is, relating to organisations below state level, such as local authorities.

as network output overestimated by as much as 50 %! This clearly complicates, or renders impossible, the realisation of contractual objectives, and then leads to a revision of investment programmes, which increases expenses in an already tense financial context.

Successes and failures in developing countries also teach us that existing systems of governance must be coherent with the private sector when utilising its services. In particular, it is necessary to give a private sector operator enough time to restore an initially dilapidated set-up, and enough autonomy to benefit to the greatest extent from the expertise they bring. Reducing the room for manoeuvre given to the private operator in the initial contract disables delegated management and restricts its scope. The systems of governance must also promote staff training and skills transfer. That is a priority for the future of emerging countries on the road towards modernising their economy.

• The findings of the recent World Bank studies

The in-depth study published by the World Bank in 2009[12] sheds important light on the debate in developing countries. For the first time, a major effort has been made to gather and analyze indicators of the performance of public-private partnerships, providing a basis for an objective assessment of their impact in terms of greater access to water and sanitation, improved quality of service and operational efficiency, as well as pricing. Some figures to start with: in 2000, around 100 million people in 43 developing countries were served by private operators; the number had grown to 160 million in 2007. Just 8 % of the public-private partnerships implemented since 1990 have been cancelled before term. The highest failure rate was in sub-Saharan Africa, where one partnership in two has been wound up. This is also the continent were the legal and financial security surrounding contracts was weakest.

Leaving aside a handful of high profile failures, the overall picture shows that services outsourced to private operators have performed satisfactorily. This is especially so in terms of improved quality of service, reduced network leakage, and payment recovery. Since 1990, more than 24 million people have gained access to public drinking water systems thanks to a private operator. In Buenos Aires, the termination of the concession should not distract attention from the quality of what was achieved, particularly in terms of the number of new connections.

As to the impact on prices, this is a red herring. Private operators are often brought in as part of a wide-ranging reform aimed at improving the

----

12. Public Private Partnerships for Urban Water Utilities: a Review of Experiences in Developing Countries.

technical and financial viability of the water sector. When the initial price is far below operating costs, then it is inevitable that it will rise, regardless of whether the service is provided by a public or a private operator. The World Bank published another study in 2008[13]. This was an econometric study of nearly a thousand public and private water distribution companies worldwide. Overall, it found no statistically significant difference between prices charged by private operators and those charged by public ones.

Beyond these considerations, public-private partnerships appear to be a viable option for developing countries, provided they are properly designed, particularly as regards the apportionment of risks and responsibilities between the parties. The main contribution a serious private operator can make lies in improved quality and more efficient service, not in the provision of direct financing. Many of the best performing contracts are those where a private operator assumes the operational and commercial risks, but not the major capital expenditures. These need to be financed out of public and private funds, depending on the particular circumstances. In practice, The World Bank recommends three approaches for effective public-private partnerships. Firstly, it invites them to be realistic, to take the time necessary to understand the local reality, without trying to sort everything out during the first years of the contract, time to train staff to adapt to their business culture, and time for the partners simply to get to know each other. Secondly, the Bank recommends assimilating what a partnership means, collective work, and ongoing day-by-day cooperation to resolve difficulties. It is not a question of the contractor handing over responsibility for the service, and "offloading" a complex problem onto a third party, but of taking on the political responsibility. Success or failure will be a joint achievement for both partners. Lastly, the Bank recommends correct evaluation, taking local conditions into account and defining a contract appropriate to the situation, both in its term and in its provisions. For The World Bank, the successes in Colombia, Morocco and West Africa came mainly from the suitability of the contracts to the local realities.

• Private operators, and the extension of a drinking water service to all

In poor countries, can a private operator participate in the reduction of the number of people without access to drinking water and sanita-

---

13. Does Private Sector Participation Improve Performance in Electricity and Water Distribution?

tion? In its 2006 Global Human Development Report, the UNDP refocuses the debate on the central criterion, upon which all those concerned with water provision should agree, in saying: "Many publicly owned utilities are failing the poor, combining inefficiency and unaccountability in management with inequity in financing and pricing. [...] The criterion for assessing policy should not be public or private but performance or nonperformance for the poor." The UNDP also adds that: "the debate on privatisation has sometimes distracted attention from the pressing issue of public utility reform." It calls for the creation of professional operators, of whatever status, capable of using a sustainable and cheap technology in order to bring drinking water to those who have not got it.

On this criterion set by the UNDP – "progress or lack of progress in favour of the poor" – private operators have nothing to be ashamed of in their work. In sub-Saharan Africa, despite the difficulties faced, 20 % of the population connected over the last few years, were connected by an individual connection provided by a private operator. The fact that more than 24 million people have gained access to drinking water thanks to public-private partnerships in the developing countries over the past 15 years is by no means negligeable, given that only a small proportion of these countries' water and waste water treatment services are privately operated.

The frequently passionate debate on the role of private operators in the water management of these countries is subject to an ideological seesaw, divorced from any deep analysis of the reality. There were those, funders in particular, advocating "privatise everything" in the name of structural adjustment. In the case of a service as sensitive as water, that theory inevitably led to the failure of any external intervention that was suffered rather than desired. There was "small is beautiful", where the funders favoured local or regional operators to the detriment of international groups. This principle showed its limitations in the great cities with millions of inhabitants. Then "Water Operators' Partnerships" (WOPs) came into fashion. When the Veolia Environnement Foundation accompanied GRET (Groupe de recherche et d'échanges technologiques), an NGO, to set up small private operators in Cambodia, it was creating a WOP, without realising it. When Veolia Water associated with the Grameen Bank to prevent the rural inhabitants of Bangladesh being poisoned by the arsenic contained in their country's groundwater, it was also creating a WOP. Nevertheless water is not provided to those without it by acronyms, but by what the acronyms represent: the ongoing struggle for effectiveness and sustainability, the will, and the generosity to provide a good water service.

IV. WATER: IS THE SERVICE TOO EXPENSIVE?

WATER IS NOT OIL

In a parallel with black gold, water is sometimes called blue gold. It is nothing of the sort: water is not oil. Oil is a fossil resource, exploited by drilling; water is a renewable resource. According to the calculations of Global Water Intelligence, the world average price of water charged to the customer is 75 cents per m^3 of water in areas covered by a distributor, as against \$320 per m^3 for oil! This study dates from two years ago when the oil price had not reached the heights it attained in 2008. Such a price differential leads to radically different economic systems with, on the one hand, a product that is so expensive that it is transported from one end of the Earth to the other, and on the other, a low-price, heavy product, which therefore has to be a local service. There is a single world market for oil, and an atomised market for water. The oil economy is unstable, at the mercy of the ups and downs of international relations and the law of supply and demand, whereas there are hundreds of thousands of local water services, regulated by the public authority, which most of the time is also local.

Electricity is not a good comparison, either. Water is a product that is easy to store but difficult to transport over long distances. Indeed water is the exact opposite of electricity, which is simple and cheap to transport, but very difficult to store. The electricity produced by the Itaipu dam, built on the Rio Paraná, on the frontier between Brazil and Paraguay, supplies the São Paulo region 1,000 km away. Such cases are extremely rare with water. Towns that have to resort to it usually do so because of the dryness of their climate. This is the case with Riyadh, that brings in sea water from 400 km away and desalinates it. Electricity plants rarely get their primary materials (oil, coal, uranium) from the site where they stand, but bring it in from a distance. They serve a regional, rather than local, market and may even export their production.

This remains the exception in the case of water, which tends to be locally managed by catchment area. It is not the great river basins that determine the division into districts for water provision, but these local catchment areas. The water sources are normally available on the spot, where the water is consumed. The boundaries of this catchment area will depend upon both geography and economics: they end where drinking water supply pipes become too expensive in terms of investment and running costs, and this is the level at which water resources are best administered. This is also the best level at which to manage pollution, almost at source, preventing rivers carrying pollutants on to villages and towns further downstream.

THE PRICE OF WATER IN FRANCE: THE PERCEPTION AND THE REALITY

In France the total water bill is divided between drinking water production and supply, waste water collection and treatment, taxes and charges payable to various public bodies. The bill usually consists of two parts: the standing charge (a fixed charge for a given period) and usage (calculated according to a meter reading or an estimate). Hence the price of water breaks down as follows (according to 2007 data):
– Drinking water production and supply represents 45 % of the bill. This item covers all the operations necessary to supply the water: sourcing it, making it drinkable, transporting it, analysing it, depreciation and operating costs of plants, maintenance of pipework, customer service management.
– Collection and treatment of waste water represents 37 % of the bill. This item covers all the operations linked to the collection of waste water and treating it before returning it to the natural environment, as well as depreciation and maintenance of the sanitation plants.
– Taxes and charges payable to public bodies represent the remaining 18 % of the bill. Two charges, called "preservation of water resources" and "fight against pollution" are paid to the Water Agencies. A third charge is payable to a French public body VNF: (Voies navigables de France) when the water is sourced from rivers, streams or canals officially considered to be navigable. Finally, VAT is added to all the items at the rate of 5.5 %, except in prescribed particular cases.

Today there is a lot of debate about the price of water, particularly in France. Some people think it is too high and that water management lacks the elementary transparency that is expected from a public service. What is the truth of the matter? The water bill represents an average of €1.00 per day per family for 330 litres of water delivered and then purified daily[1] Households spend 0.8 % of their average budget on water and sanitation, that is, three times less than they spend on

1. *Les services collectifs d'eau et d'assainissement en France*, report by BIPE/FP2E, January 2008.

14. Breakdown of water charges in 1996 and in 2007

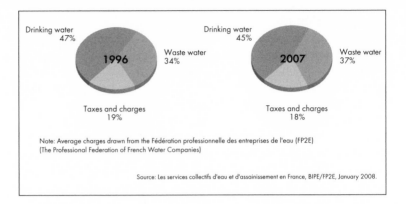

Drinking water 47%
Waste water 34%
1996
Taxes and charges 19%

Drinking water 45%
Waste water 37%
2007
Taxes and charges 18%

Note: Average charges drawn from the Fédération professionnelle des entreprises de l'eau (FP2E)
(The Professional Federation of French Water Companies)

Source: Les services collectifs d'eau et d'assainissement en France, BIPE/FP2E, January 2008.

telecommunications (2.4 %), and over four times less than on electricity (3.8 %)[2]. In other words, water is one of the cheapest public services. If we take a look at the international scene, we see that, according to NUS Consulting[3], the average price of water and sanitation services in large French towns was €3.01 (including taxes) per m^3 on 1 January 2008, which is below the European average of €3.40 per m^3. Despite this, the price of water is seen as too high. According to the annual SOFRES-C.I. Eau survey, 60 % of French people think water is "rather expensive", but an even higher percentage (64 %) do not know how much it costs. Not actually knowing the price of water does not in any way prevent people from forming the opinion that it is too high! Those who venture to quote the price per m^3 overestimate it quite sharply: the average they give is €4.80, which is 59 % higher than INSEE's average price.

Criticism is also directed at any increase in water price. Why does it increase? In France, as elsewhere, the water service is becoming more and more complicated. Taking into account the multiplication of sources of pollution and stricter environmental legislation, it is no longer just a question of ensuring access to drinking water and sanitation, but much more broadly, ensuring that the entire water cycle is managed as well as possible. In this respect, environmental questions and the fight against pollution give rise to extra costs for communities and water professionals. For water supply and sanitation, as in many other sectors, standards have stiffened. What was acceptable yesterday is no longer

----

2. Source: INSEE (National Institute for Statistics and Economic Studies), 2006.
3. The NUS Consulting study outlines each year the development of water prices in large European cities. It studies the price of water in the five largest cities of ten European countries (Germany, Belgium, Denmark, Spain, France, Finland, Italy, Holland, UK and Sweden). The 2008 study is based on prices for January 2008, for 120 m^3 per year. The study is carried out at the request of FP2E.

acceptable today. After a period of significant price rises in the water service, with increases between 3.7 % and 8 % per year, caused by increased community investment (in particular for purification of waste water in order to apply the 1991 European Directive), the increase in water pricing has slowed down and stabilised since 1999. If we examine the percentage of income that French people allot to water, we can see that it is unchanged since 1996.

Comparison with neighbouring European countries is instructive. Since the launch of the NUS Consulting study in 2003, between July 2003 and January 2008 the average price of water in France has increased on average by 3.6 % per year. Over the same period, for Europe, it has increased annually by 5.1. %.

15. Trends in expenditure on water in household budgets

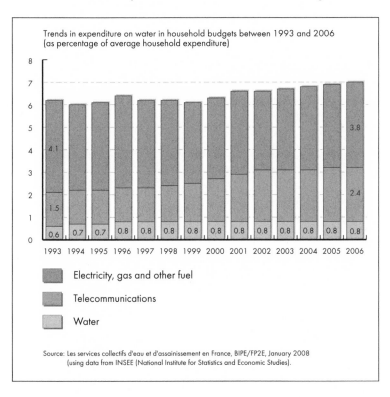

Trends in expenditure on water in household budgets between 1993 and 2006 (as percentage of average household expenditure)

Source: Les services collectifs d'eau et d'assainissement en France, BIPE/FP2E, January 2008 (using data from INSEE (National Institute for Statistics and Economic Studies).

WHAT PRICE FOR WHAT AREA? THE EXAMPLE OF FRANCE

A recurring question preoccupies many French people. Why are there price differences between different communities? Let us pause to look at the facts. France's 36,700 communes are grouped into 12,400 water services and 16,700 sanitation services. The elected representatives in charge of each service fix its price in accordance with local technical conditions, such as the state of the untreated water resource, the quality of the service provided, or the advance in sanitation programmes. The quality and quantity of the resource is very unequal in different territories, and that is the reason for different costs and the consequently different price to communities. The differences may be stark, but according to IFEN, the French Environment Institute, 80 % of communities fall within a ratio of about 1 to 3, that is a price per m³ of between €1.30 and €4.02. In fact, tariff disparity between the principal towns is less in France than in most other European countries.

16. The prices of water in European countries in 2008

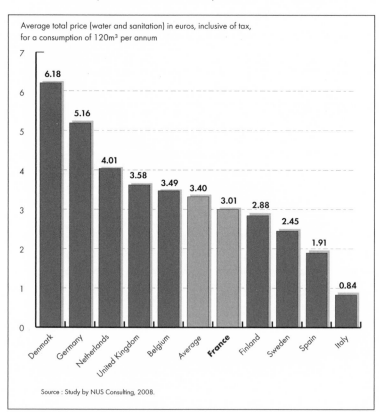

Average total price (water and sanitation) in euros, inclusive of tax, for a consumption of 120m³ per annum

Source : Study by NUS Consulting, 2008.

Price differences give rise to questions about price harmonisation. This forces us to ask which territory should act as the standard, or, in other words, what is the right level at which to set collective rates? That, in turn, takes us back to the key question of setting parameters to define good management of water services.

First of all, there is a technical constraint: the water service's primary material is local and its finished product, drinking water, is also local in the sense that it cannot be transported over a long distance at an affordable price. As a result, the technical infrastructures for production and distribution of water services are also local and serve a limited territory. So a sensible principle to adopt is that operational and technical coverage should match as closely as possible. Water service management is the better for covering, as far as possible the same area as the technical network. The elected local representative is a much better manager than the regional or national one, who, by definition, is further away. Water is a local service and should stay that way, as a matter of economic efficiency. So its price should be set in accordance with the local technical and operational parameters. Local management, with a local price for a local service. But consumers resent price differences between communities in the same area, so it is often necessary to harmonise the price of water in conurbations or single areas, which is justified to the extent that at this level, technical systems are increasingly unified. Many authorities have already done, or are doing this.

Should the price of water be standardised on a larger scale, at regional or country level? Setting a single price at this level would mean doubly reinforcing irresponsibility. Firstly, it can reward collective irresponsibility, because towns and communities with less efficient management would have no incentive to improve. Good water service managers would pay for the bad ones. If complete cross-subsidisation of prices were established, residents who had already paid to raise their sanitation infrastructures to meet standards, would have to pay again for communities who had not made the effort to do so. But it would also be rewarding individual irresponsibility, because the community would have to finance questionable decisions taken by some of its members, such as the construction of buildings a long way from the existing network. A community cannot guarantee the right to water at the same price to those who want to live out of the way, unilaterally exposing the community to additional costs.

Everything leads us to believe that in France cross-subsidisation of the water price is a utopian idea. It is one of those idealist theories that some people would like to impose on the water industry, following the pattern of quite different sectors such as electricity or gas. By doing so, they are ignoring the geographical and practical conditions that apply to the supply of water. They are implicitly demanding a decrease in the responsibilities of local representatives. They are calling for the creation of a blind, centralised system, which will cause waste and lack of responsibility on the part of consumers, such as local communities. In general, local representatives are against extended cross-subsidisation for the price of water, because it would penalise well-managed services and would not encourage others to make the necessary efforts to improve. However at the level of conurbations or single local areas, cross-subsidisation makes sense. Harmonising the water price at this level meets one of the expectations of consumers who live in a homogeneous area. As the 2001 Tavernier report[4] indicated, "although a single price [for water] is unimaginable, the price jungle is unacceptable."

ARE PUBLIC OPERATORS LESS EXPENSIVE?
A FALSE EVIDENCE

A study carried out in 2006 by the Boston Consulting Group for the Federation of French Water Enterprises, explains the difficulties in finding the appropriate indicators with which to compare water prices in France. In fact, "when the operational context is identical, if we limit ourselves solely to the price on the bill, delegated management appears to be more expensive – by between 5.5 % and 9.5 % – than direct public management. But in the identical operational context, if the calculation takes fuller account of the taxes paid by delegated management and the fact that direct public managements sometimes finance operations via taxes and not by billing – the total cost of delegation of the public service is seen to be significantly lower– from 3 % to 7 % lower – than that of direct public managements."

Moreover, when comparing costs, it is essential to remember the fact that operational conditions are far more difficult, and therefore more expensive, for delegated managements because it is these more difficult situations which have often induced communities to delegate the service to an experienced professional. It has already been pointed out that delegated managements usually have to deal with poorer quality untreated water than direct public managements, since they have, on average, half the available groundwater. They manage two and half times more coastal communities than do public operators. These coastal communities are characterised by larger installations to cope

---

4. Information report on the financing and management of water, presented by Yves Tavernier. A French MP on 22 May 2001.

17. Immediate changes in the price of water
after coming under public management

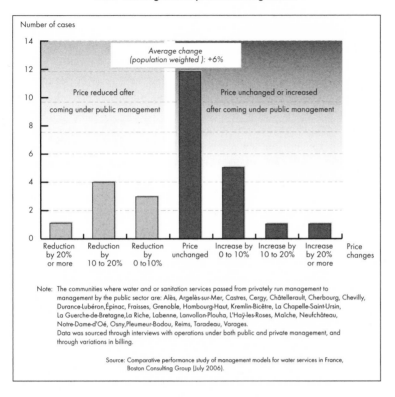

Number of cases

Average change
(population weighted): +6%

Price reduced after
coming under public management

Price unchanged or increased
after coming under public management

| Reduction by 20% or more | Reduction by 10 to 20% | Reduction by 0 to 10% | Price unchanged | Increase by 0 to 10% | Increase by 10 to 20% | Increase by 20% or more | Price changes |

Note: The communities where water and or sanitation services passed from privately run management to
management by the public sector are: Alès, Argelès-sur-Mer, Castres, Cergy, Châtellerault, Cherbourg, Chevilly,
Durance-Lubéron,Épinac, Fraisses, Grenoble, Hombourg-Haut, Kremlin-Bicêtre, La Chapelle-Saint-Ursin,
La Guerche-de-Bretagne,La Riche, Labenne, Lanvollon-Plouha, L'Haÿ-les-Roses, Maîche, Neufchâteau,
Notre-Dame-d'Oé, Osny,Pleumeur-Bodou, Reims, Taradeau, Varages.
Data was sourced through interviews with operations under both public and private management, and
through variations in billing.

Source: Comparative performance study of management models for water services in France,
Boston Consulting Group (July 2006).

with peak tourist periods; they have poorer resources, because they
are situated at the point of convergence of rivers and pollution; and they
have stricter standards for waste water disposal, because they are
near bathing areas. All these factors increase the costs of their water
services. Lastly, "just 15 % of the water dealt with by delegated managements needs only simple treatment (according to the DGS[5] classification), as opposed to 37 % of the water for services under direct public
management." So, of course, that gives rise to different operational
costs.

Although cases are rare, it is interesting to measure the price
changes when a service is returned to direct public management. In the
majority of cases (19 out of 27), the water tariffs remained unchanged or
were increased. And after returning to direct public management, the

----

5. DGS: Direction générale de la santé. According to the DGS classification "simple" is category A1, "complete" comprises categories A2 and A3, and "mixed treatment" means a mix-
ture of waters needing different treatment. Source IFEN, INRA-LERNA.

prices increased during the following years by an average of 3.8 % per year. So it is not true that return to direct public management brings about a systematic price decrease.

What happens in other countries? France, which has the highest level of delegated management in Europe, has a water price below the European average. Countries like Denmark and Germany, where private operators are not often used, have a price higher than €5.00 per m³, that is, €2.00 higher than the French average. England has gone for the fullest intervention of the private sector, in the form of privatisation, rather than delegated management: it has a water price slightly higher than the European average, but lower than Holland, where there are no private operators in the supply of drinking water and very few in sanitation. In Spain, there are as many people served by a private operator than by a public operator: this country has one of the lowest prices in Europe, which is half that of France, but Spain heavily subsidises its water from taxes.

Although when we include the capital works done, for example, the construction of new infrastructure, resorting to the private sector may lead to a water price increase, private sector intervention is not synonymous with a higher price. In 2002, The Hague, capital of Holland, commissioned a private company to build and finance its new waste water treatment plant, to modernise the one that already existed, and to run both of them. In order to maximise quality and reduce costs, the invitation to tender required the candidates to propose a price at least 10.5 % lower than that obtained by public management of the project. The Defluent group of companies came up with 17 % economies in relation to the public price and was taken on by the authority, the Delfland Water Board. In the USA, the contract to design, build and run the Tampa Bay drinking water plant brought $85 million worth of economies to the delegating community, thanks to the technical know-how and economies of scale brought by the private sector.

We are bound to conclude that a private company's need to make a profit does not lead to higher water prices than those set by public operators. The economies, from which the local authority and the population benefits through good management of its infrastructures and cost-control, together with a better quality of service, are much more important factors than the profits made by the private company.

Many public services, including transport, health, electricity and motorways throughout the world, use the cooperation of private enterprises, which, like all their equivalents, have to make a profit. The same goes for water. There is therefore no incompatibility between the existence of profit, indispensable to the sustainability of a private enterprise, and the provision of a quality public service. Through their professional

WATER

WATER'S FALSE FRIENDS

efficiency, their organisation and their results-based contracts binding them to the public authority, private businesses compete to provide the best quality public service. The profit the private company legitimately expects from doing its job is part of the price of the package it offers. Nevertheless, profits made from water arouse great resentment, whereas that does not happen, or happens much less, in other sectors.

If it is the very existence of profit that is condemned, then the question must immediately be asked: must improvement in the water service be banned, although all will benefit from it (the community and its residents), just because the private operator will also make a profit from it? If it is the rate of profit made by private companies that is being criticised, does that mean that it is too high? In France the average profit margin made by water companies is 6 % of turnover after taxes, and that is a reasonable return in comparison with other public service activities. These profits are below the figures for EDF (Electricité de France), Paris Airport, and for French motorways. Is it common knowledge that in Holland public water companies make considerable profits, up to nearly 40 % of turnover for some of them? Execrated when it is made by private operators, should profit be "exonerated" when it is made by public operators? If it is considered just to reward public shareholders, why is it not just to reward private shareholders? If, however, profits made by private water companies are regarded as unjustified because of the way they are subsequently disbursed, let us look at that. They are used for investment, particularly in new infrastructures intended to improve the water and sanitation services; to finance research programmes in order to develop tomorrow's technologies; to pay dividends to shareholders; and to be returned in part to employees as a form of worker profit-sharing.

A business that does not make a profit is a business that is not investing and not doing research. And it is not an accident that in France, as a study by BIPE emphasises, the greatest research and development effort for water and sanitation services is made by the private sector[6].

A PRICE THAT FAILS TO REFLECT THE SCARCITY OF WATER

Although the quantity of water available is finite, its price does not reflect its scarcity. Angel Gurría, OECD Secretary General, forcefully reminded us in May 2007 that the world needs higher water prices. In many countries, the price for water is generally set too low. This is the case in many communities in Italy. Despite endemic drought affecting

---

6. BIPE Study, *Eléments pour un benchmark des services d'eau et d'assainissement*, 2003.

several of its regions, Spain maintains its system of subsidy, especially for irrigation, and charges one of the lowest prices in Europe for water. In certain towns situated in the heart of the Sahara, water is free. That is without even mentioning water for agricultural use, which is usually supplied at a symbolic or zero rate. We should not delude ourselves: there is a direct link between overexploitation of the resource and under-pricing. Prices that are far too low perpetuate the illusion that water is super-abundant and that nothing is lost when it is wasted. As the UNDP put it in its 2006 Global Human Development Report, "If someone were selling Porsches for three thousand dollars a piece, there would be a shortage of those too." Likewise, in a context of growing scarcity, one of the main goals of tariff policies must be to put a price on nature and a cost on pollution.

Although it is desirable to make payment for water to give a better indication of its scarcity in its price, that does not in any way reduce the need to be sure that everybody, even the poorest, has access to this essential service. As we have already made clear, a social tariff can be set, like those in operation in very many countries, or a system of direct aid as in Chile: there mechanisms for assistance combine with sea-sonal water tariffs to reflect water's scarcity better. "Whenever scarcity is not dealt with by economic mechanisms, in addition to scarcity, there is sure to be wastage of the remaining resources[7]." Ultimately, scarcity results from supply and demand. But the two parts of the equation, supply and demand, are affected by political choices. The relation between supply and demand depends on water price or on maximum quotas, which, everywhere in the world, are set by the public authorities.

If we want water to be really sustainable, we must recognise its occasional scarcity and to set up mechanisms to manage it. But do we really want to do this? Here there are many believers, but few who want to practise what they preach... We know very well that water is at a crossroads where everything meets. When public authorities fix the price of water or put a limit on its use, they have to arbitrate between numerous legitimate but conflicting interests: environmental protection, the financial stability of the service, but also economic development, public health, ensuring that the water service is available to the poor, etc. Supplying water is a complex activity, and so is managing it sustainably.

----

7. Olivier Godard, Director of Research at CNRS, 2006.

THE DEVELOPING WORLD: UNDER-FINANCING
CAUSES EXCLUSION

Water is always too expensive for the poor who are not connected to the public network. They pay ten to twenty times more per m³ than those who have access to the public water service. They also have additional penalties: a poor service since they do not have running water at home; more diseases, because they get worse quality water, with discontinuities in the supply; less school for their children because of the time they spend going to fetch precious water for their families. For those who are connected to the public network, however, the average water price is usually too low. In 2002, according to a study by the National Institute for Urban Affairs, most Indian towns charged an average water price equal to one tenth of their costs for operation and maintenance[8].

Even in the higher price blocks for water, the bills sent to customers do not cover the costs, and this was true for cities in India's neighbouring countries, such as Dhaka and Kathmandu. A fundamentally unfair subsidy emerges: because they use more water, households with higher or medium incomes receive higher subsidies, and, of course, lower income households receive lower subsidies because they consume less water. Although the situation has changed somewhat, even today very few Indian towns charge more than a quarter of the costs they incur for operation and maintenance.

In 2003, the Asian Development Bank assessed the relation between potential receipts and actual receipts for ten water companies. The calculations were made respecting the criterion of water price affordability, so that households on an income of less than $5,000 per year would not spend more than 5 % on their water supply. The figures speak for themselves: Phnom Penh could double its average water price; Colombo and Shanghai could triple it; Dhaka could quadruple it; in Vientiane, it could be multiplied by six, and in Delhi by nearly ten!

We also see a refusal by the public authorities to adjust their tariffs over the years to take inflation into account. According to a study published in June 2005 and carried out in five countries of South East Asia (Indonesia, Malaysia, Philippines, Thailand, Vietnam) only Thailand periodically adjusted its tariffs. In Malaysia, one in two water services had not raised their tariffs for ten years.[9] The same thing happens in Africa[10]: the ratio between the actual cost of water and the average price charged by the public services frequently lies between four and ten.

----

8. Raghupathi and Foster, *op. cit.*
9. Source: USAID – US-AEP – *Full cost recovery in South East Asia*, Hanoi, June 2005.
10. Bertrand Dardenne, *op. cit.*

18. Potential and real revenue from water services in Asia

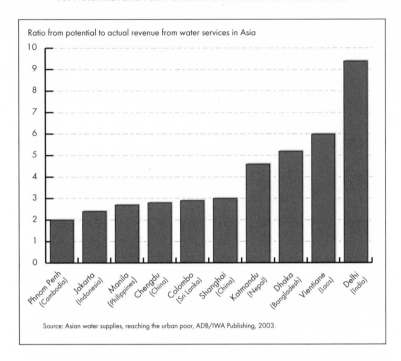

Ratio from potential to actual revenue from water services in Asia

Source: Asian water supplies, reaching the urban poor, ADB/IWA Publishing, 2003.

In the end, maintenance budgets are reduced, water services do not have the necessary financial resources to renew their capital, service quality declines and networks are not extended to those who are not connected. Another consequence is loss of customers' trust and respect for the public water service. Indeed, throughout the world, many water services are locked into a cycle of under-financing, under-maintenance and poor quality. And what is worse, not fixing the level of drinking water at the right price means condemning those who do not have any, to never have any.

INFORMATION AND TRANSPARENCY: AT THE HEART OF THE DEBATE

The debate about price comparisons raises the more general question of information about the performance of water services. As water has now become everybody's affair, the demand for information has grown. Many efforts have been made by the different players to respond to this demand. In France, how is access to information and the debate in the world of water managed? The proceedings of the Water Agencies,

WATER

WATER'S FALSE FRIENDS

mostly composed of elected representatives and customer delegates, ensure transparency at the level of the local water area. The local public services advisory boards, compulsory since 2003 in communities with more than 10,000 inhabitants, strengthen citizen participation in the provision of services. Results of water analyses can be consulted by all in every town hall. As for finance, the principle "the user pays for water" guarantees the publishing of its price and a better understanding of all the costs.

When a local authority decides to delegate its service, it can strictly control the private operator. Annual technical and financial reports, audits and regular meetings punctuate the relation between the public authority and its contractor. In 2003 the National Audit Office recognised the progress made in transparency in France: "Information for elected representatives and users has greatly improved with the generalisation of annual reports… and the standardisation of bills." In relation to other sectors, public service delegations offer a seldom equalled transparency. Not many other businesses are obliged to set out their costs and margins so fully to their customers! When an individual or a company buys a service, the supplier is not obliged to reveal its profits and the full details of its costs. But this is not the case when a town entrusts the management of its water service to a private company.

Moreover, the private companies took the initiative quite a few years ago now to create a system of performance indicators to improve information about the public services whose management has been delegated to them. In France, this step was recognised in the 2007 decree on the mayoral reports on the price of water and service quality, which provides for the deployment of performance indicator systems, not only for private businesses, but also for public operators.

But there is still one point to be considered. The decree published in France in 2007 concerning mayoral reports on the price of water and service quality, does not in fact settle the question of communicating these data to the public. Performance evaluation should be accompanied by wider communications to ensure transparency and to foster debate and dialogue, particularly within communities, Water Agencies and local public service consultative committees.

All these efforts can be seen and recognised: over ten years the percentage of people who say they are not well informed enough about water has dropped by 15 points. However, nearly two in three residents still want more information. And whatever progress has already been made, those involved in water must do more. They must increase their efforts where the public demands it: according to the Water Information Centre, the public's three main expectations concern quality control of drinking water, standards for its distribution and lastly, where it comes from. Price only comes fourth[11]. They must also respond to residents' concerns, in particular their fears about scarcity and pollution of sources by nitrates and pesticides. That means being more careful about water availability in summer, and supervising and remedying the quality of water resources, as the 2000 European Directive recommends.

We are now in a world where the quality of information about water is as important as the quality of water itself. Transparency means trustworthiness and availability of information. If the information is independent and unbiased, in the necessary debate on the efficiency of water services, comparisons will enable people to replace passion with reason. Now is the time for those in the water industry to provide themselves with the tools to assess more calmly the real performance of services. No fully credible comparison can be made without objectively evaluating performance and without an autonomous body supplying information that is independent of the operators (both public and private). It is up to the public authority to define such a body. This triple process of evaluation-information-dialogue which, by the way, can apply to all water services, in all countries, whatever their degree of development, is rarely carried through to the end. But it is the price of transparency.

KNOWING WHAT IS BEING COMPARED

Ignoring some of the facts, such as the relative performance of operators, constitutes a denial of information, which leads to a distortion in debates. As bad money drives away good, so false debates drive away the true. The democratic debate about water is more than legitimate, as long as it is seriously based. So it is important to be rigorous when comparing different water services. These must always be analysed taking into account the local context, service performance and conformity with standards. Otherwise, comparison of water prices will be biased, and can act as an invitation to demagogy. A water service that does not respect the regulations in force may offer a price today that is lower than its neighbours, but tomorrow it will have to resort to investments, which will raise the price. There is no point in comparing water prices without checking whether the maintenance and renewal policy is appropriate to the state of the plant and networks, or whether costs are being piled up for the next generation.

----

11. Source: C.I. Eau [Water Information Centre]/TNS [Taylor Nelson Sofres] SOFRES [Société française d'enquêtes par sondage].

V. FUNDERS, OFTEN CRITICISED
BUT INDISPENSABLE

International financial institutions are not loved. They are accused of being "accomplices" in world economic liberalisation and of having imposed painful structural adjustment programmes on countries on the verge of bankruptcy. Some of their solutions are considered to be inappropriate in dealing with a region's economic collapse. Indeed, the 1997-1998 financial crisis in South East Asia, then in other emerging countries, has shown that several countries, which came out of it best, followed the opposite strategy to the measures recommended by the International Monetary Fund (IMF). International financial institutions are blamed because, sadly, the remedies they propose are not painless and affect poorer populations. They are criticised for not reflecting the differences between countries in their administrative advice, and for depending too heavily on the developed countries that founded them or endowed them with capital. Their high running costs and their battalions of experts shock the poor. They are decried for interfering in the management of developing countries or, on the contrary, for their passivity. They are reviled for doing too much or too little. They are criticised as much for their power as for their impotence. Finally, one of their great wrongs is not to have achieved a goal that depends but little on them: banishing wretched poverty from the world.

In the case of water, international financial institutions have been taken to task for having encouraged, in developing countries as elsewhere, full cost recovery from service users or, at least, fuller recovery. International financial institutions have also encouraged public-private partnerships, which are disliked by a good number of associations. In the past they recommended privatisation, including the sale of infrastructures to private operators, which was clearly inappropriate for developing countries. For water as for other sectors, they prefer to give loans, rather than grants. In theory, loans encourage recipients to made good use of the moneys received, but also forces them to repay, which is obviously disagreeable. Who does not prefer a gift to a loan, even at subsidised rates? Because of the way in which international financial institutions function, most of their financial aid is given to states, and they cannot pay them directly to local authorities or to NGOs (even if this constraint has been wiped out, as, for example, when aid is conditional upon results). This results in a partial siphoning off of the funds by intermediary bureaucracies, and sudden disappointments when one calculates the sums that actually reach projects on the ground.

Together with development aid agencies and other funders, international financial institutions suffer from being organisationally bloated in the multiple structures dealing with water questions. At UN level alone, 23 agencies or programmes are involved with water. If the funders recognise the vital importance of water and sanitation in human development, this priority does not translate in amounts lent or given. Their reputation suffers from slow progress, which remains well behind expectations. Delays in advancing towards the Millennium Goals, for instance, have raised questions about their efforts and efficiency. Although many programmes they have financed have improved water and sanitation services, many others have not achieved a service of sustainable quality, and many advances have been simply obliterated by the frantic growth of towns in many developing countries.

Nevertheless, international financial institutions, development aid agencies and other funders in general, have a key role to play in access to water. This role is not only concerned with finance and innovation, but also with their contribution to the debate: they move it forward through the initiatives they fund and the ensuing evaluations.

AN OVERVIEW OF PUBLIC AID FOR WATER AND SANITATION

If we make a rapid examination of public aid for water and sanitation, what do we see? Firstly, there are not many generous countries. For a long time, the most generous country has been Japan. It is the most important source of public development aid, with an average of $850 million for the period 2003-2004, that is, a bit more than a fifth of the total aid for water and sanitation. Within the G8, Germany and Japan spend nearly 6 % of their aid on water and sanitation, as against less than 3 % for the USA, Italy and the UK.

Secondly, multilateral aid is developing. It now represents nearly a third of global aid, as against only a fifth in 1998. That is progress, because multilateral aid is directed more towards very poor countries than is bilateral aid. The World Bank and the European Union are the principal sources. Thirdly: aid remains very unevenly spread. Twenty countries receive nearly three quarters of public development aid and ten countries share two thirds. But sub-Saharan Africa, with its enormous needs, only gets a fifth. Indeed, half the aid goes towards large-scale infrastructural works, which favour urban areas to the detriment of rural areas. Funders concentrate on a small number of countries, with which they are in the habit of working, to the point of creating blatant disparities. In 2004,

19. Aid allocated for water and sanitation in 2003–2004

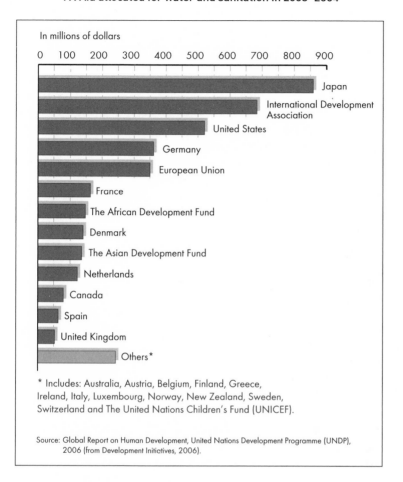

In millions of dollars

0	100	200	300	400	500	600	700	800	900

Japan

International Development Association

United States

Germany

European Union

France

The African Development Fund

Denmark

The Asian Development Fund

Netherlands

Canada

Spain

United Kingdom

Others*

* Includes: Australia, Austria, Belgium, Finland, Greece,
Ireland, Italy, Luxembourg, Norway, New Zealand, Sweden,
Switzerland and The United Nations Children's Fund (UNICEF).

Source: Global Report on Human Development, United Nations Development Programme (UNDP),
2006 (from Development Initiatives, 2006).

Tunisia and Ghana received an average of $88 worth of aid per person for water, whereas Burkina Faso and Mozambique, which are poorer, only received $2 per person. Fourthly, and most importantly, it is no surprise to anyone that aid remains far below what is needed. Water and sanitation receive about $3 billion in public development aid. However, the extra investments required to reach the Millennium Goals for water and sanitation have been calculated as between $10 billion and $30 billion per year. The minimum threshold of $10 billion corresponds to the use of "sustainable and cheap" technologies. Even if the whole of that sum cannot come from public development aid, the goal is still a very long way off.

INSUFFICIENT FINANCE

Many projects to bring drinking water and sanitation to millions of people have been accomplished with the contribution of international or national agencies. However, the work still to be done is huge, and financial efforts need to be considerably increased. A comparison will illustrate the extent of the financial challenge and the urgent need for international solidarity: connection to the drinking water network usually costs between $100 and $200 in developing countries. But a billion people in the world live on less than one dollar a day.

Financial needs and aid provision do not correspond, particularly in Africa. Water, and even more so, sanitation, are marginalised in poverty reduction programmes and public budgets. According to the UNDP, governments should devote at least 1 % of their GDP to spending on water and sanitation. In fact, they do not contribute half that amount. Finance derived from users and taxpayers exceeds public development aid, but remains very insufficient. Public development aid itself is very stingily granted: the donor countries assign less than 5 % to water. From 2000 to 2002, that is the first two years of the period fixed for reaching the Millennium Development Goals, the total sum given even decreased! We can only agree that this is an odd way of setting out to reach those Goals. Only four countries went over this 5 % threshold in their bilateral aid. They deserve to be listed: they were Denmark, Luxembourg, Germany and Japan[1].

The financial challenge can be addressed. But will that happen when we know that a growing part of public finance is directed towards coping with extraordinary natural disasters such as floods or cyclones? This gradually leaves aside the treatment of ordinary and sadly commonplace misfortunes, like the absence of drinking water, which kills ten times more people than wars. The $10 billion per year needed to reach the Millennium Goals is less than half the amount that rich countries spend per year on mineral water and less than 5 days' worth of world military expenditure. That $10 billion per year is 1/3,000 of world market capitalisation, or rather was, because this figure dates from 2007, before the 2008 financial crisis: it is only 10 % of public development aid. As the Economists' Circle and Érik Orsenna have objectively noted, it is an amount that could be described as a "planetary tip"[2]. Nevertheless, this tip is withheld.

----
1. 2006 Global Human Development Report, UNDP, *op. cit.*
2. The Economists' Circle and Érik Orsenna, *Un monde de ressources rares*, Paris, Éditions Perrin, 2007, 216 pp.

• Africa: facing the Millennium Goals

In 2005, after the Gleneagles Summit, the G8 countries committed themselves to increasing their official development aid for Africa, taking all sectors together, by $25 billion. That would amount to annual aid in the order of $100 per African. Today the G8 is lagging behind on this recent promise. In the case of water, the President of the African Development Bank, Donald Kaberuka, drew up a report in July 2008, during the 11th African Union Summit on financing water and sanitation infrastructures to reach the Millennium Goals in Africa. His diagnosis was simple: current financial flows represent only 20 % of what would be necessary to finance the Millennium Goals in Africa. On that occasion he announced that the African Development Bank was going to allocate $14.2 billion to the water sector, as one of its four priorities. One thing is certain: without a radical change of rhythm, the Millennium Goals will not be reached in Africa.

20. Financial shortfalls for the Millennium Goals

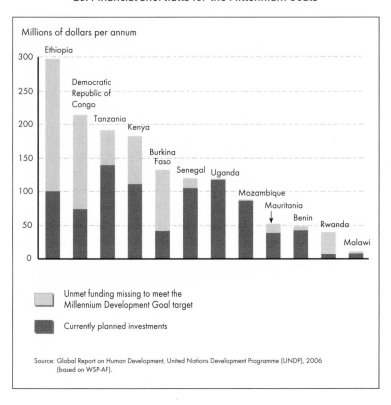

Millions of dollars per annum

Unmet funding missing to meet the Millennium Development Goal target

Currently planned investments

Source: Global Report on Human Development, United Nations Development Programme (UNDP), 2006 (based on WSP-AF).

115

• Europe relaxes its efforts

For the European Union, the 2007 figures for public development aid, with all sectors taken together, are poor. Between 2006 and 2007, aid decreased from 0.41 % to 0.38 % of GDP. Concretely, that means that €1.7 billion less were spent on projects to benefit the poor around the world. Firstly, let us recall that after the 2002 Johannesburg World Summit for Sustainable Development, the EU launched the European Water Initiative, whose aim was to make an effective contribution to the Millennium Goals in Africa. The "Platform for Water Dialogue" component has worked well and has raised awareness about sanitation problems, thanks to the advocacy of the Water Solidarity Programme. But the financial component, that is, the "European Water Facility" which was initially to be granted €1 billion, has melted away like snow in the sun. Some money has been given to NGOs or public operators (private companies are excluded) to a total of €500 million. Today, no extra grant has been programmed. Thus the "Prodi billion" has been halved. As well as disappointment about the amount, the sluggishness of procedures has also been strongly criticised by field workers. Nevertheless, as Louis Michel, European Commissioner for Development and Humanitarian Aid, forcefully declared, "We are the first generation to be able to look extreme poverty in the face and say with conviction that we have the money, we have the medicines and we have the know-how[3]." So it is essential that the European Union, international institutions and other funders continue to support water and sanitation access programmes, which cannot be carried out in towns in developing countries without a strong input from public finances.

In fact, several of the reasons explaining the lack of finance spring from the recipient nations themselves. Apart from governance, to which we shall return, we need to examine the quality of projects seeking public aid. There is probably more aid than is needed for the number of good projects, and there are certainly not enough good projects soliciting finance, in any of the recipient nations. Funders cannot be reproached for being demanding about the quality of projects to which they contribute financial resources.

THE SEARCH FOR NEW FINANCIAL INITIATIVES

International financial institutions often play a key part in the financing, and therefore in the success, of work done on the ground. They

3. "L'aide publique au développement a baissé en 2007", *Le Figaro*, 4 April 2008.

contribute some of the necessary funds but, above all, they exert leverage to attract other finance. Together with development agencies and other funders, they catalyse research efforts to invent and test new solutions, such as the following.

- From "unconditional aid" to aid dependent on results

One of The World Bank's new approaches is to link subsidy payments to actual achievement of objectives. This aid is called Output-Based-Aid (OBA), and the new model makes efficiency the determining criterion. It concentrates on the assessment of expected performances, that is, on results. It aims at better management of public expenditure and the creation of a more attractive environment for investors. So it breaks with certain traditional approaches, which take little account of results or strict use of the aid on the ground. In OBA-type financial mechanisms, the aid is only paid out once the work has been done. Consequently, they demand a very precise definition of the goals to be pursued and local pre-financing. And for those to whom it is granted, a considerable advantage is that this aid takes the form of grants, not loans.

In fact, this type of financing has existed since the turn of the Millennium, but it has only recently been applied to water and sanitation. Service providers eligible to apply for this aid may be public or private operators, NGOs, or community organisations. At the end of 2007, water represented one third of projects in the "social services and infrastructure" sector identified by Global Partnership on Output-Based-Aid. In the water and sanitation sector, projects are running in South Africa, Paraguay, Cambodia and Morocco. Depending on its results, this financial programme for the water sector could be expanded so as to contribute to the Millennium Goals. However, in its current form, this mechanism also creates a heavy administrative burden, particularly in terms of the time required for operators to assemble the necessary data for the frequent evaluations.

In 2006, AMENDIS, in charge of water, sanitation and electricity services in Tangiers and Tetuan, was selected to receive $2 million in aid for pilot OBA-type projects. There was an intrinsic interest in the experiment and its potential later repetition. Rather than going directly to AMENDIS or the city of Tangiers, this sum was made available to subsidise 3,000 poor households who were otherwise unable to pay their share for water and sanitation connection. The World Bank will repay a part, according to the number of households connected to the drinking water and sanitation network. So it is the private operator and the public authority who bear the financial risks of this operation, before being reimbursed, at least partly, if the goals are reached.

• Towards a doubling of French public aid for water services?

Water is one of the priorities for French public development aid and France is one of the chief funders in this sector. Between 2001 and 2003, France contributed an average of €268 million per year in bilateral aid and €100 million per year in multilateral aid. After the Evian G8 summit in June 2003, France announced the doubling of its support to the water sector. This commitment was a step towards reaching a global development aid effort in 2012 corresponding to 0.7 % of GDP, as recommended by the UN.

This direction followed the policy of the French Development Agency (AFD). Water is one of the main sectors where this financial institution intervenes. Its strategy turns on three axes: resource management, access for all to a drinking water and sanitation service, and thirdly, water and agriculture. Water represents 20 % of the AFD's commitments. From 2001 to 2005, the average sum of its commitments in the water sector was raised to €145 million per year. The principal destination was Africa, which receives two thirds[4]. The AFD have estimated that during this period more than 20 million people have benefited from its financing of water and sanitation. In order to double the French aid contribution for water, it anticipates that its annual aid will reach €290 million in 2009. Sanitation will represent a growing part of this sum, reaching 40 % in 2009, then 50 % in 2011, that is, parity with drinking water.

The AFD contributes to the leverage effect, such that, where sufficiently mature decentralised cooperation projects exist in communities in developing and emerging countries, it can bring in complementary funding. In this way, the projects expand and are replicated.

• New forms of association

A further example of a new form of financial intervention comes from the World Bank and the French Development Agency. In December 2007, the International Finance Corporation (IFC), The World Bank's private sector arm, and AFD subsidiary PROPARCO (Société de promotion et de participation pour la coopération economique) invested in Veolia Water AMI (Africa Middle-East India) as part of a capital increase. IFC and Proparco took 13.89 % and 5.56 % respectively of the capital of Veolia Water AMI. The aim was to support the development of infrastructures in these three areas of the world. The new shareholders

----

4. Cadre d'intervention sectoriel 2007-2009, Agence française de développement (AFD), Conseil de surveillance, 29 March 2007.

contributed to this operator's development projects by putting in their own funds. Additionally, because both are serious heavyweights, they facilitated access to other funders. As Rashazd Kaldany, IFC head of the infrastructures department, stresses: "There are enormous needs to be met and access to local financing is primordial to obtaining permanent investments in water infrastructure." This institutional partnership with recognised players means that greater importance will be given to social and environmental needs. It will enable the development of projects in terms of environmental management and the attainment of the Millennium Goals, as well as assuring better financial security and better governance.

OBA financing, the increase of public development aid that some countries devote to water, the entry of international institutions into operator capital, Water Operators' Partnerships (WOPs) and the resources associated with them are initiatives that need to be increased, especially in Africa. It is important to serve poor populations by means of these original partnerships, which combine the professionalism of experienced operators and the legitimacy of internationally recognised funders. Access to water in developing countries, particularly sub-Saharan Africa, can only be achieved through cooperative, creative models. Public and private players together with civil society, can invent practical solutions, away from sterile ideological confrontations that sap energy and foster inaction; and away from passing fads which tend to focus on only one way of operating to the detriment of others.

WHEN ONE METHOD DISPLACES ANOTHER

During the 1990s, the majority of international institutions, particularly The World Bank and the IMF, called for greater involvement by the private sector in the supply of drinking water and sanitation. Unsurprisingly, that aroused the wrath of opponents of delegated water management. Passing from one "magic formula" to another, these same institutions are today all in favour of partnerships between two operators. NGOs and trade unions struggled to ensure that these forms of cooperation – which in the jargon of water professionals are known as Water Operators' Partnerships (WOPs) – are exclusively reserved to public operators, but failed. WOPs are supported by the UN Secretary General's Advisory Board, and they function on the basis of bilateral cooperation between several players. They are extremely flexible, both in the way they work and in the nature of their partners.

These new types of partnerships will only work if they last, which means unfailing and long-term political support, and if they achieve a real transfer of skills and if they focus on results. So should we, as appears to be the case, make these Water Operators' Partnerships "the" solution to the problems of water access? After having favoured public-private partnerships, don't international institutions risk causing new disappointments if they turn these WOPs into a panacea?

Indeed there is no single overall solution. No set formula ("all public", "all private", or "all NGO") and no cooperation model (whether it be "public-private partnerships", "operator-operator partnership" or "private-public partnership") can solve the water question alone, or attract sufficient financial resources to satisfy the enormous needs. First and foremost, water access in developing countries requires pragmatism. It needs effective solutions that suit particular local situations. Before being set up as a model, every initiative requires numerous experiments and rigorous assessment. Then we can avoid the "canonisation" at international conferences of models about which those on the ground remain sceptical. International institutions must show the way, and must not hesitate to test different models. They must also continue to support reflection by rigorously assessing each initiative, so as to learn all the possible lessons.

PART 3

FINDING NEW MODELS

The challenges facing global water management are many: tension over resources, changes to local hydrological cycles caused by global warming, delays in providing sanitation, chronic pollution of freshwater resources, effective implementation of the right to water and sanitation for poor communities, the emergence of new types of pollution such as pharmaceutical residues, and so on. But solutions to the problems do exist, and they can be deployed as soon as the conditions for good governance are met.

Water supply is a political issue in the best sense of the term. Because of the need to reconcile different imperatives, water management is a laboratory of democracy, that is to say, it is a paradigm of the search for compromise between legitimate but diverging interests. Managing water means first of all choosing between different constraints and opportunities, whether environmental, social or economic. Unfortunately there are no perfect solutions, only a limited range of optimal solutions, a range we must constantly seek to widen. The most "revolutionary" approaches to water management are often utopian, and are sometimes designed just to win political popularity. As the sociologist and philosopher, Edgar Morin, said, "Giving up the best of all worlds does not mean giving up a better world[1]." •••

1. Edgar Morin, *Éduquer pour l'ère planétaire. La pensée complexe comme méthode d'apprentisssage dans l'erreur et l'incertitude humaines*, Balland, 2003.

••• Our responsibility tells us to go beyond fatuous and over-simpli-fied slogans such as "free water for all", so that we can concentrate on elaborating models that are truly capable of answering today's issues. There is something wonderful about working with water, in that it makes us feel we are carrying out a worthwhile task. We are forced to see things in a new light if we wish to benefit from water while at the same time safeguarding it. The work also promoted a greater mutual understanding; ultimately, it appeals to everyone's good sense and intelligence.

In this respect one point is central: the economic model underlying transactions over water is not pernicious. More than ever, water needs an economic and financial system which can organise its distribution intelligently and guide its management towards goals established by local and national communities. Whatever detractors may say, the economy is without doubt a servant of water. It is a silent, not to say austere, servant, a mechanism of progressive adjustment between different interests, whether consumers, public bodies, citizens, share-holders, operators or associations: an extremely difficult exercise. "Free water for all" registers much higher in terms of popularity, but the most popular policies are not always the most effective ones. Our challenge, then, is not to set public service against delegated or private management, but to adapt the economic and financial model for water so that public and private operators can coexist. Continuous innovation is needed to ensure that the economy remains an efficient servant for all, and does not become an anonymous and uncontrollable master.

The economic ecosystem in which water services exist is constantly changing, and the services have to adapt their way of operating to each new order. At the beginning of the twenty-first century, we need models capable of taking up the numerous challenges. To do this, we must re-examine not only the technological and financial principles behind the way models operate, but also the way they fit into political systems and society. Today, as always, "The story of water management is at once a story of human ingenuity and human frailty[2]."

There are several forms of productive intervention in water services:

– Exploitation of new resources. Globally, demographic growth is inevitable and populations will be concentrated more and more into urban areas, i.e. in narrow strips of the Earth's surface. In the context of a growing need for water, the use of new resources cannot be ignored, whether these are classic resources, such as water that has leaked from distribution systems, or alternative resources such as recycled waste water.

– New ways of paying for water and sanitation services. Additional tasks are often imposed on water services that act to increase their costs. Water conservation policies often reduce the income of a service with more-or-less fixed costs. There are clearly financial limits for many water services in developing countries. All these factors lead public authorities to look for ways of changing the economics of water services. They must create economic models capable of financing and maintaining the infrastructure, of adhering to stricter regulations, of providing safe drinking water and sanitation for all, and of giving water services the means to improve customer management and thus making constant progress.

– New partnerships. The constantly increasing environmental demands, and the need to bring safe drinking water and sanitation to those who lack it in developing countries, demand that water services develop new methods of cooperation. This includes new ways of relating to research organisations, local associations, atypical financial institutions (like the Grameen Bank) and the many participants in the day-to-day economy.

– Increased social cohesion. This is necessary in developing countries to bring safe drinking water and sanitation services to all who lack

----
2. Human Development Report 2006, UNDP, *op.cit.*

them, but it is also needed in developed countries so that people in marginal situations are able to maintain access to these two services.

New water resources to meet new shortages. New income to finance new services. New collaborations to bring together the necessary skills. The increase in social assistance to make it easier for all to have access to water services. New techniques for dealing with new types of pollution and limiting their impact on the environment. This could be the route map for safe drinking water and sanitation services at the start of this twenty-first century, but it would be incomplete without adding governance. Without progress in this, we cannot hope that water resources will be adequately protected or that one day everyone will have access to safe drinking water. Without good governance, there can be no usefully applied technology, no good management of water services, and no preservation of the environment.

I. NEW RESOURCES

WATER SAVING, A RESOURCE AVAILABLE NOW

Avoiding leaks and putting an end to our consumerist relationship with nature are imperative when water and energy resources are both diminishing. In many cities in America, Asia or Africa, more than 40 % of water is lost through pipe leakages. In the distribution network in Colombo, Delhi, New Orleans or even Riyadh, leaks account for 50 %. In other words, for every two cubic metres of water taken from the natural environment and treated, one cubic metre disappears while it is being transported to the final consumer. In France, the average rate is around 20 %[1]. There is a threshold, of course, beyond which the cost of increasing the proportion of water delivered by the system becomes unreasonable, but we cannot continue to allow leaks equivalent to half the volume of water entering the distribution system.

The reduction of leaks will clearly allow a reduced rate of abstraction from resources, or enable greater needs to be satisfied with the same rate of abstraction. This measure should precede all research into new sources. In fact, saving water in the public system is often the most substantial water resource immediately available. Indeed, conserving water first of all implies efficient delivery. A good public service operator must conserve the scarce resources in his care, and then create new ones.

It is possible to recover enormous volumes of water by preventing loss from urban networks. For example, the reduction in losses over five years in Tangiers, Tetuan and Rabat, is equivalent to the amount consumed by nearly 800,000 inhabitants. Already in November 2004, Ali Fassi-Fihri, director general of the Moroccan Office nationale de l'eau potable (ONEP), jokingly reminded people that the battle against leaks in the water systems of Moroccan cities had, in one year, allowed a saving of 5 % of treated water. This was bad news for the finances of ONEP (which sells treated water to the distributors), but excellent news for an arid country like Morocco.

In Paris, over the last 30 years, the yield from the distribution system has risen from 75 % to 90 %, says Odile de Korner, acting director general

----

1. Source: *Les dossiers de l'IFEN – Dossier no.7:* "Les services d'eau en 2004", Oct. 2007.

of Eau de Paris, a semi-public company retained by the Ville de Paris[2]. This period corresponds more or less to that of the management of water distribution by private operators. On the right bank of the city, where Veolia Water manages distribution to 1.4 million Parisians, the network yield has increased spectacularly since 1985 (the beginning of the leasing contract) and reached 96.2 % in 2007. This improvement has allowed a reduction of nearly 400 million m^3 in the volume of water pumped from rivers and water courses since the transition to delegated management, which is nearly four times the annual consumption figure for the right bank area. This has produced marked savings in investment, whether in the sourcing, pumping, treatment or transport; savings which benefit both the city and its inhabitants.

• Moving from a culture of supply management to demand management

Managing demand saves water. Until now, supply management has predominated over demand management, but it is now crucial to involve consumers more so that they can take control of their consumption and protect their environment. But this "civic responsibility towards water", that is, the adoption of behaviour which promotes the general interest, will not increase unless inhabitants are given the means by which to control their consumption. This can be dome by the widespread installation of household water meters and remote meter-reading or text-message information systems. In Shenzhen, in China, 70,000 text messages are sent every month to consumers, including some that alert them when their consumption is greater than usual. In Paris, Metz, and other areas covered by the Syndicat des eaux d'Ile de France, remote meter-reading systems are gradually being installed. Thanks to these, consumers can monitor their water consumption in real time and thus take greater control over it.

Nor does industry wish to be outdone. Between 2000 and 2007, Danone reduced its water consumption by 30 %, and in the last ten years, L'Oreal has lowered its consumption by 23.5 %[3]. By a combination of giving more responsibility to employees, recycling waste water and systematic research into water saving, they have put in place, or caused to be put in place, methods that are more economical of water.

2. Fabienne Lemarchand, "Le long périple de l'eau de Paris", *La Recherche*, July-Aug. 2008.
3. Source: "Les Groupes du CAC 40 tentent de réduire leurs consommations d'eau", *La Tribune*, 29 Aug. 2008.

• Why encourage individuals to save water?

We need to specify the reasons why consumers are asked to save water. When there is a shortage, it is obvious, but we need to understand a number of implications. First of all, people should not save water to the point of taking risks with hygiene or their health. The injunction to "turn your tap off firmly" does not carry the same weight as economising on a non-renewable resource such as oil. Except for water pumped from underground fossil water, water is continuously renewing itself. This act, then, cannot then be compared to turning out the lights as one leaves the office in order to burn less oil. Nor is the object to save water in your home in order to solve the problem of shortages elsewhere. Since transporting water over long distances is not conceivable, water savings made in Europe do not help Saharan Africa to access larger quantities of water.

The true logic of water saving by individuals lies in the optimisation of the water cycle. Water wasted because of a tap that has not been turned off properly or leaks from a toilet cistern will sooner or later be returned to the system, to be turned once more into drinking water, but that can take a long time. When water has become scarce, fighting losses from public and private systems increases availability and this is fundamental if needs are to be satisfied in arid regions. In places where water is abundant, it is still necessary to fight leakage, as this reduces costs to both community and consumer. Treatment of water and waste water carries costs, so running drinking water through an additional cycle without having used it, generates unnecessary costs.

• Conserving water in the fields

Agriculture is the largest consumer of water globally, and it is also the largest waster of water. It uses more than two thirds of the water taken from the natural environment, greatly outstripping both industry and individual users. It is true to say that the topics of agricultural water and food security cannot be separated, and the latter is a permanent preoccupation of many African and Asian countries. Choose whether to drink or eat! This deliberately exaggerated phrase nevertheless contains a grain of truth: the necessary increase in agricultural productivity cannot be achieved without a less wasteful use of "field water".

How can we get agriculture to use less water? There are two major points to consider. First of all, a coherent strategy cannot be put in place without first deconstructing the existing incoherence, that is to say, without giving up policies which reward waste. At the top of the list are the perverse subsidies which amount to real encouragement to exploit underground water sources. Next, and at the same time, micro-irrigation must be expanded. In a conventional and extensive spray-irrigation system, two thirds of the water used does not reach the plant. Drip systems, which take the water right to where it is needed by the plant, allow the water used to be reduced by two-thirds. In Jordan, the development of drip-irrigation systems has reduced national agricultural water consumption to a third of what it was before. But Jordan remains an exception; globally, this technique is used on only one per cent of irrigated land. This is not to advocate a policy of discouraging irrigation, but one of making irrigation more "abstemious" in its use of water. After all, without artificial irrigation systems, harvests would be reduced by 40 % globally. Only when the battle against wastage has been truly joined should political decision-makers begin to think about finding new water resources for agriculture.

RECYCLED WASTE WATER: THE ONLY RESOURCE THAT GAINS FROM ECONOMIC DEVELOPMENT

In emerging countries, growth in demand for water accompanies economic development, just as it did in Europe and North America in the nineteenth century. This is a good thing. The growth in water consumption points to better hygiene, a reduction in disease, better quality of life and businesses which contribute to an economic boom in the country. We cannot at the same time call for development of poor countries and a reduction in water consumption by their inhabitants. Our grandparents took a bath once a week. We take a bath or shower every day, which multiplies water consumption by seven. Nobody claims that this is not progress. Should we then criticise the rise in water consumption in developing countries? Should we not rather be happy about it?

Of course, this sort of development also fuels tensions over water resources if they are scarce. But where there is not enough water, the solution lies less in "sharing the scarcity" than in the use of alternative resources. Two techniques are currently gaining in popularity, as they open up access to non-conventional resources: waste water recycling and sea water desalination. These processes have been in use for centuries, but technical advances over the last few decades have greatly enlarged their application.

• Water is too precious to be used only once

During the twentieth century, many forms of scarcity have been turned upside-down: new scarcities have appeared and old ones have faded away. Shortages create breakdowns in our economy, our relationship with nature and our way of life. However, new scarcities invite us to create new resources. Thus waste water has now gained the status of a resource. While what we once thought we had in surplus has now become less abundant, what we used to regard as waste has been transformed into a resource. Waste water, that used to be a dangerous nuisance, is now judged to be useful.

Waste water recycling is a tried and tested method of producing water for industrial, agricultural or domestic purposes. The purified waste is collected as it leaves the treatment plant and then receives additional treatment. This varies according to the use to which it will be put, whether it be industrial cooling water, irrigation of fields, garden watering, refilling underground water courses, and so on. Whatever the use, waste water re-treatment installations demand a professional approach and constant reliability, without which the health of whole populations could be put at risk.

Waste water recycling, which, in turn, may be multiplied by the creation of many urban mini-cycles, prevents water being returned to nature after only one use. This artificial short-circuit in the natural water cycle does in one short stage the purification work that nature does over a long journey that carries river water to the sea, then from the sea to the clouds, and finally to the rain that falls on the land. Even better, recycling reduces the amount of purified waste water returned to the environment and, in so doing, helps to break the all too common link between urban growth and pollution of aquatic environments.

Recycling waste water is without any doubt a promising path capable of providing large volumes of water. It is also anticipated that in the next decade installations will be built which will quadruple world capacity for recycling; all the more so as the waste water is located exactly where it is needed. It is also the only resource that increases with economic development, in parallel with the growth in need for water. Waste water recycling prevents or lessens conflicts over use, which are inherent in the growing water demand. In financial terms, it is a way for the community to save. Indeed, in arid regions, recycled water is cheaper than desalinated or imported water. This technology is a double winner; by circulating recycled waste water straight into the local water cycle, it creates a new resource, which will allow more "sustainable" consumption of water, while at the same time it reduces the environmental impact of sanitation.

At the present time, Israel is the foremost user of this technology, with three quarters of its waste water being reused for irrigation. But Windhoek, the capital of Namibia, has gone the farthest in terms of quality of recycled water. This city re-treats its waste water, converting it directly into drinking water for its inhabitants. It is true that Namibia has the unenviable privilege of being the most arid country in southern Africa. The nearest permanent river to Windhoek is 600 km away. This city can only rely on local resources by using them in the best possible way. In 2001, it had a new facility built to recycle waste water for domestic use: the New Goreangab plant supplies nearly 300,000 inhabitants. In order to control health risks, the process includes many barriers against pathogens: pre-ozonation, coagulation/flocculation, flotation, rapid sand filtration, ozonation, filtration, activated carbon adsorption, ultrafiltration and chlorination. This facility allows the capital to face up to its chronic supply deficit, and without it there would be a shortfall of between 30 and 35 % in the city's water resources. Only two cities in the world, Windhoek and Singapore, practise large-scale waste water recycling to produce drinking water for their inhabitants, but in the city-state of Singapore, recycled waste water counts for only a minor percentage of the volume of water returned to drinking water reservoirs, as against more than one third in Windhoek.

Several countries are rapidly strengthening their waste water recycling capacity, especially Australia, which is struggling against drought. Faced with very scarce rainfall and uninterrupted growth in the demand for water, it is mobilising alternative resources. Southern Australia, the capital of which is Adelaide, is the most arid state of the most arid continent. In the fullness of time, the public authorities hope to reuse 50 % of waste water in order to have available water resources that are less dependent on climatic variation. The Bolivar recycling plant plays a key role in this programme. It can re-treat 43,000 m^3 of water per day to irrigate 200 farms, recharge the water course and store water there for pumping during the summer, and finally provide water for garden watering in residential areas. Farther north, the state of Queensland has launched the Western Corridor project, the largest waste water recycling project in the southern hemisphere, which has a production capacity of up to 232,000 m^3 per day.

• The psychological constraints on the use of recycled waste water

Objections to recycling come less from the technical aspects than from the psychological. "Recycling" is understood as "reuse of rubbish". Despite its potential, waste water recycling remains an unacceptable solution in some cultures, even for irrigation. Strong psychological barriers must be crossed if re-treated waste water is to become acceptable, and this challenge is far from being won.

In 2006, in Toowoomba, an Australian city with 90,000 inhabitants, the mayor organised a referendum asking people for their opinion before starting to use recycled waste water. The people rejected this project, even though it had been accepted by the health authorities. The continuing drought may well prompt them to reconsider this decision. The inhabitants of Windhoek had no choice but to get as much as they could out of their extremely scarce water resources and they have backed this system for 40 years. There, the success of waste water recycling for human consumption rests on the reliability of the technology, public information and the absence of any economically viable alternative. In France, the plan for managing water scarcity launched by the government in 2005 depends in part on waste water reuse. According to a CECOP survey carried out in 2006, 90 % of the French population support recycling for irrigation and garden watering.

SEA WATER, AN UNLIMITED RESOURCE

Sea water is the most abundant resource on Earth, and represents 97.5 % of the planet's water reserves, yet globally scarcely one per cent of drinking water is produced by desalination. On the other hand, 40 % of the world's population live less than 70 km from the sea, that is, in an area where desalination could reasonably be carried out, if this technique is chosen for serious development.

In the Middle East, in many of the Caribbean and Polynesian islands, around the Mediterranean rim, in Australia, life would be impossible without sea water desalination. Saudi Arabia draws 70 % of its water from the Persian Gulf and the Red Sea. Spain has backed desalination in answer to the rise in demand for water and its limited freshwater resources and is at the top of the league in Europe in terms of volume of desalinated water. China, India and the United States are getting ready to launch large desalination projects and even England is showing interest in this solution for London. In ten years or so, the capacity for global sea water desalination production should have doubled: more and more plants will be "drinking up" the sea in order to satisfy human needs. Desalination will allow us to avail ourselves of, effectively, an unlimited alternative water resource.

In southern Israel, the Ashkelon region is facing a severe water shortage. In order to satisfy the growing need for water, the local authorities decided to draw supplies from the sea and the Ashkelon plant opened in December 2005. It is the world's largest plant using membrane desalination. It produces enough water for 1.4 million people and supplies 15 % of the country's drinking water. The cost of the drinking water produced at the Ashkelon plant is one of the lowest in the world for membrane desalination of sea water, with a factory-gate price of €0.5 per m³. This performance can be explained by good energy efficiency, economies of scale and the continual reduction in the price of membranes, which shows that membrane desalination is an economically competitive option.

Critics of sea water desalination have focused their attention on its high energy consumption, maintaining that it is not relevant at a time of climate change and that it endangers the environment. It would be absurd to claim that desalination has no impact on the environment, but it seems unreasonable to exclude this process when it brings water to populations which would otherwise face shortages – even more so when hoped-for technological advances promise to reduce the amount of energy needed to make it work.

As for waste discharged into the sea by desalination plants, this presents two difficulties, temperature and chemical composition – neither of which is insuperable. In thermal processes, the discharges are between 5° C and 15° C warmer than sea water, while the brine produced by the other major technology which is used, reverse osmosis, has a salt concentration double that of sea water. In addition, the waste water discharged into the sea from both processes contains some chemical pollutants: chlorine in the case of thermal desalination plants, or anti-scaling agents in reverse osmosis installations. Nevertheless, dilution and dispersion technology, along with a careful choice of discharge location, can prevent chemical or thermal imbalance in marine ecosystems.

At a time when nations are rediscovering the benefits of energy independence, it is no bad thing to remind ourselves of the advantages of "water independence", too. Sea water desalination, like waste water recycling, reinforces a country's autonomy in water supply and allows it to reduce or to avoid water imports from abroad. It lessens tensions between states, which could be fuelled by scarcity of water. It gives access to a source of supply that is reliable, independent of unpredictable rainfall and based at home, where it is sheltered from international constraints. Singapore is very aware of this, as it counts on desalination and waste water recycling to free itself from the need to import water from Malaysia.

• Energy: an ecological and financial challenge

Energy costs are high on the list of expenditure for water and sanitation services, and even more so when their raw material is purified waste water or sea water. In increasing order of energy consumption, first we have recycled waste water, then sea water desalination by reverse osmosis, and finally thermal desalination. Energy consumption was 20 kW per m³ of desalinated sea water in 1970; it has now fallen to less than 3 kW for the most energy-efficient processes. At the same time, the cost of membrane desalination has dropped considerably, so that this technology is no longer the preserve of rich countries. Even if great progress has been made, the cost of desalination remains heavily dependent on energy consumption. To give just one example, the membrane desalination plant at Carboneras, in Andalusia, provides 120,000 m³ per day, but it consumes a third of all the electricity in the province[4].

Whichever process is used, it is essential to monitor its energy efficiency in order to control costs and minimise the discharge of greenhouse gases. The tendency of oil prices to rise over the long term makes this issue even more pressing, especially in regions where cheap energy is not available, such as islands, whose isolation tends to increase the costs of energy supply.

Membrane technology is energy hungry, but it opens up access to alternative water resources, whether from recycled waste water or from desalinated sea water. Some countries and communities run the risk of being caught in a dilemma, if water independence can only be achieved at the cost of greater energy dependence. In order to help them confront this situation, current research aims to lower the energy consumption of membrane processes even more. It also aims to supply membrane desalination facilities with electricity produced in part from renewable sources. As for new thermal desalination projects, desalination plants are increasingly being installed in conjunction with energy production installations. The heat produced when hydrocarbons are burnt to produce electricity is used to vaporise sea water. These hybrid solutions allow optimal use of thermal power stations.

• Alternative sources as an answer to climate change

Forecasters tell us to expect greater numbers of extreme climatic events. Increasingly irregular hydrological cycles will force communities to strengthen the security of their drinking water supply. Many arid regions are already hostages to water. They risk becoming even more so as climate change approaches, altering the distribution of water in both space and time. Hydrological maps will show radical redistribution between countries and regions of the world. In southern Europe, for instance, the gap between growing needs and diminishing resources is certain to grow. Conversely, northern Europe is likely to

----
4. Sabine Lattemann, "Le dessalement est-il écologique?", *La Recherche*, July-Aug. 2008.

see increased rainfall. Glaciers containing huge reserves of water are gradually melting all around the world. As they retreat, they will release less water during the dry season and thus fail to maintain river levels.

"Security through diversity" is the policy implemented by various cities of Australian, a country facing its eighth consecutive drought year. The vast programmes of sea water desalination and waste water recycling aim at building a water supply system independent of erratic rainfall. Other countries are redrawing their hydrological maps by instigating huge projects to transfer water between river basins. China has started work on diverting the River Yangtze for thousands of kilometres, in order to supply Beijing and its surrounding area. India envisages linking the Ganges and the Brahmaputra as part of a vast programme that aims to interconnect 46 rivers through 30 canals with a total length of 10,000 km. In 2008, Brazil relaunched an old project to divert the São Francisco over 500 km in order to irrigate the parched northeast. In France, a project to transfer water between the Rhone and Catalonia, in Spain, which has been on the agenda for a long time, has been abandoned for the present.

In extreme cases, saving water, storing it, transporting it or producing it from alternative sources may still not be enough, if the roots of the problem have not been tackled. In Australia, in the Murray Basin, irrigated agriculture draws 80 % of the available water flow. This forces us to ask what is the basic problem? Is there too little water? Or are there rather more cotton plantations and more animals than the environment can support? The long-term solution may be to change the economic model of the region.

USING PREVIOUSLY UNEXPLOITED SOURCES

The appeal of these alternative sources should not mask the interest in exploiting entirely new freshwater sources, at least where this is possible without causing environmental damage. Indeed, this is the source of water that services have experience in treating and consumers are used to using. It does not cause any psychological blocks like recycled waste water and even when it is of poor quality, treatment costs are considerably less than for sea water desalination and, usually, recycled waste water. There is however a precondition to drawing on new freshwater resources: their rate of flow must be correctly evaluated and no more water withdrawn from them than will be renewed. This condition seems obvious, but it has rarely been respected, as can be seen in the general overexploitation of underground water sources in several parts of the world.

Clearly, once rivers finish their journey to the sea what is not abstracted from them is lost. This is why the protests that emerged when six round trips of tanker ships were organised by the Société des eaux de Marseille, in spring 2008, to transport water to Barcelona, appear bizarre. At that time, the capital city of Catalonia was suffering from severe water shortages. The emergency water came from the Rhone, which was on high flow. This water was drawn only a few hundred metres upstream from the sea, where it would have mixed, in any case, with sea water. As announced at the time, this operation was a temporary one. These few localised and time-limited transfers were never going to unbalance the hydrological flow of the most powerful river in France. All things considered, there is far less need to make a fuss over such emergency transfers of water than over the policies that, through lack of foresight, made them inevitable.

• How realistic are new freshwater resources?

While greater exploitation of existing resources is out of the question in many arid regions affected by decreased rainfall, in other places this is the solution of choice. France, for instance, has an annual precipitation of 480 billion m^3, of which 300 billion are returned to the atmosphere through evaporation and plant transpiration. The balance reaches 175 billion m^3 when account is taken of water imports from and exports to neighbouring countries via rivers. Out of this mass, around 100 billion m^3 soak through the soil and feed into underground water sources, which in turn help to refill the rivers. The rest runs off after rainfall and flows directly into streams and rivers without passing through underground watercourses. What proportion of these resources do the French use? According to the Bureau de recherche géologique et minière (BRGM), "We draw 30 to 40 billion m^3 annually, of which the majority is returned to aquifers, and 8 billion m^3 are abstracted from underground watercourses[5]." France has more than 500 underground watercourses, of which 400 cover more than 10 km^2. The amount of potentially useable water stored in these watercourses is as much as 2,000 billion m^3, but only the renewable proportion of this should be exploited, i.e. 100 billion m^3 per year. Compared with the eight billion m^3 that are currently being taken, it leaves a considerable margin to expand the exploitation of underground resources!

However, taking more fresh water from aquifers is only possible where they are not overexploited. Since 2003, many of them have been

----

5. Quotation from Thierry Pointet, hydrogeologist at the Bureau de recherche géologique et minière (BRGM), in "Les nappes phréatiques remises à flot", le Figaro Magazine, 3 May 2008.

struggling to regain their normal levels. Heavy rainfall in winter 2007 and spring 2008 refilled some of the large underground watercourses, but many remain underfilled, for example in the departments of Poitou-Charentes, Aude and Pyrénées-Orientales.

• Refilling aquifers: a weapon against drought

"In order to compensate for water shortages, we need to create a global storage capacity of 200 km³ between now and 2025," explains Yann Moreau of the Berlin Centre of Competence for Water, an international research centre specialising in water studies. To this end, we can either build new dams or refill groundwater sources. Groundwater sources are effectively water reserves. They play the same role as a dam, but below ground. This is an essential consideration in hot, arid regions, where there is heavy evaporation from large expanses of surface water, sometimes amounting to 50 %. Storing water underground prevents these evaporation losses. Refilling groundwater resources allows the water to be produced at any time, especially during droughts, to satisfy the many needs of agriculture, industry and domestic users. In Australia, Texas, Florida, California and Berlin, in Germany, groundwater sources are currently being refilled in order to increase the availability of water resources. In the context of growing scarcity, refilling groundwater sources will be used more and more to help to ensure the supply of drinking water.

When overexploited aquifers are artificially recharged, they can be restored and brought back to their original level. In the case of coastal groundwater, refilling them with fresh water prevents sea water penetration and salinisation. But it is possible to go farther than this. Refilling a groundwater source boosts its potential: it becomes more productive than it would be naturally, so this technique can form part of a strategy for active management of groundwater resources.

RAINWATER: A USEFUL RESOURCE, BUT NOT WITHOUT ITS RISKS

This overview would not be complete if it did not mention rainwater collection, an ancient practice in arid regions. In the Greek islands, the inhabitants still collect rainwater to satisfy their water needs. In Rajasthan, people store rainwater then use it throughout the year. In towns situated right in desert, like Jaisalmer, families collect every drop of rain that falls on their roof or in the surrounding area. In Kenya, rainwater collection tanks are so precious that they are sometimes included as part of a young woman's marriage dowry! Rainwater collection is also practised in some countries with a heavy rainfall, such as Cambodia, because they also have a prolonged dry season, which forces people who are not connected to mains water to store water for themselves.

In Europe, rainwater collection for domestic or industrial uses is becoming very popular. In northern Europe it is practised widely, and the practice is beginning to spread in France. Collected water, usually from roofs, is stored in a tank instead of running straight into the sewage or rainwater systems. It is generally used for watering parks and gardens, cleaning paved areas, roads and railways, or washing vehicles.

Rainwater collection is in fact the logical counterpart of waste water recycling. If water from sewage networks can be collected and re-treated, why not do the same for rainwater? Making use of rainwater clearly shortens the water cycle and creates a direct supply exactly where it is needed. For private users, the simplest, the least dangerous and the most ecological method of collecting rainwater to water the garden is to use a water butt. Rainwater can also be used by industry for part of a production process. At its Maubeuge factory, Renault has installed a rainwater recycling system, which supplies the vehicle assembly line. In 1999, before this system was in place, as much as 570,000 m³ of water had to be bought, while the run-off, over the 39 hectares of impermeable surface, came to around 300,000 m³ per year, so by collecting this water a large proportion of the needs of the site could be met. Aéroports de Paris, which also owns a vast impermeable surface at Orly, recovers rainwater and directs it back into their air-conditioning systems.

Conversely, experience has shown that directing rainwater inside homes has not been a good idea. Bringing it inside the home introduces germs, a veritable nutrient broth, close to vulnerable people such as children and the elderly. There is also the chance that faulty plumbing could inadvertently connect the rainwater system to the drinking water system, making the risk of polluting the drinking water system difficult to control. It is too easy to forget that rainwater is not necessarily drinkable, especially if it has been stagnating for several days in a tank (even more so if the tank is not cleaned regularly, as is usually the case). During the symposium organised on 16 November 2006 by the Cercle français de l'eau on the topic "Considered water management: towards another culture?", Daniel Yon, from France Nature Environnement, said, "I am, for reasons of health security, completely opposed to the introduction of rainwater inside homes." Prudently, the French authorities, by a decree of 21 August 2008, authorised the use of rainwater for washing laundry, only "as an experiment".

The results of experiments on using rainwater inside homes in European countries are clearly negative. A detailed study on this topic was carried out in the Netherlands in 2006. What were the results? Faulty plumbing was common, and seemed impossible to avoid, as the systems in private houses are frequently modified for this purpose. There were alarming numbers of cases of gastroenteritis due to contamination of drinking water with collected water. Many people used their untreated rainwater to fill their children's paddling pools, despite repeatedly having been warned not to. Already in 2003, Swiss authorities had pointed out that as domestic rainwater reuse systems had only a minimal ecological impact, it would be better to favour water-saving systems. For its part, the German government calculated that the costs of reinstating the drinking water system after accidental pollution by rainwater would be very heavy: in some cases more than several hundred thousand euros. So, taking into account the real health risks involved, these countries strengthened their legislation or recommendations on the use of rainwater, so that it cannot be brought into homes without thorough treatment, and the cost of this makes it more expensive than tap water.

II. NEW ECONOMIC AND FINANCIAL MODELS

Technical solutions and new sources of water, however efficient and promising they may seem, will not be enough to overcome all the challenges facing global water management. Economic models also need to be changed in response to new objectives, sometimes contradicting the old ones. We need new tools to deal with and to adapt better to new constraints in water management. Our present model reflects the preoccupation with "hygiene" inherited from the beginning of the twentieth century. Public health is historically linked to the development of the trades and professions connected with water, and to their economic model. It was in the general interest to increase household consumption of water in order to make noticeable improvements in hygiene. The economic model as it was then defined – and which we still use today – was in line with this preoccupation with public health. It aimed to encourage operators to increase the volume of water consumed. Water services were paid according to the volume of water sold, while their costs were mainly fixed. This model has largely reached its objectives: the health revolution enjoyed by the great cities of Europe in the nineteenth century, and which in fifty years added ten years to average life expectancy, was closely linked to the collection of waste water via drains and to the supply of safe drinking water.

Now, though, several factors lead us to consider revising this economic positioning of water services. First, the increasing scarcity of freshwater resources runs completely counter to the economic logic of water services, which drives the operator to increase the volume of water consumed so as to be paid more. Next, in most towns and cities in developed countries, a pincer effect between cost increases due to regulations and the steady decline in the volume of water charged for undermines the financial stability of the services, leaving no option but to raise the price per unit of water, which many elected representatives refuse to do. In France, for example, consumption fell by 4 % in 2007. On the other hand, people now expect much more from the water services than they used to. They have been burdened with additional tasks, which may include rainwater treatment, cooperation with other agencies, improvements on roads and railways, river maintenance, etc. These additional tasks increase expenditure without any adjustment having been made in the economic equation. Finally, like other public activities, financing water services has a systematic problem. It is an industry with fixed costs, whose product, water, is charged for by volume; 80 % of its costs are fixed while 80 % of its receipts are variable. This payment method undermines the financial stability of the service when water consumption drops. While gains in productivity partly compensate for the lack of income, they have mainly come from private operators, and they can only be taken so far. This raises another paradox: the more water-saving policies bear fruit, the more prices per unit must go up to cover the costs.

Thinking of new models is not a minor issue. We must build an economic structure, and consequently a payment system which is not out of kilter with the general interest of the community, nor with the cost structure of water services. The recent necessity to limit quantities abstracted from nature overturns the financial logic for water companies: far from seeking to sell more, the water service must now seek to sell less, even though its income still comes from sales! In order to reconcile the environment with water service finance, the water service must be paid in such a way as to conserve water resources, but without the danger of sacrificing health benefits to environmental protection. Already, the growing scarcity of water has led many services to give more weight to informing and teaching the public. They invite their users to reduce their water consumption. But they have to go farther, to revise, and even reconstruct whole sections of the economics of water. No economic activity can last if it runs counter to the long-term interests of the community and the land where it is carried out. The issue then is to rethink the economic equation of water services while at the same time retaining its public health objectives.

MIXED FUNDING BETWEEN SERVICE USERS
AND TAXPAYERS

There are several different ways that the economic model for water could be re-established in France and in some other developed countries. The first would be to evolve towards mixed financing of water services, based on both taxpayers and users, and not solely on the latter. For the service provider, more and more responsibilities now diverge from the supply of drinking water and sanitation in the strict sense, such as management of rainwater and flood defences, and the restoration of water courses and their surrounding area. These benefit all the inhabitants of a region and should be paid for by them all, not solely by water service users. There would be an evolution from financing in which 100 % is borne by the user, to one in which the user would pay, for example 70 %

and the taxpayer 30 %. Other ancillary services which are charged to the water service would need to be identified, the costs calculated and the charges paid by those who benefit from the services. Separating receipts from the number of cubic meters billed would, on the one hand, make the revenues less variable, and on the other, would lessen the mismatch between the receipts structure and the costs structure.

Mixed financing systems are already in place in the Netherlands and in Italy, but they do not finance additional functions of the water service, only its original role. What we are proposing is radically different. The objective is to make the user of water services pay for all those elements that are genuinely related to the drinking water and sanitation services, while the rest is paid by the taxpayer. In other words, it is about going back to a strict principle of "the user pays for water", which has otherwise been left farther and farther behind. Many emerging countries have the opposite problem. There too, steps must be taken to apply the "user pays for water" principle better, but by raising the tariffs so that they will cover water service costs better, especially costs relating to keeping the system working. In these countries, too few of the operating charges have been integrated into the user's bill, while in developed countries like France, the user has been billed for too many additional expenses, not all truly relating to drinking water supply and sanitation.

At the same time, a better way must be found to have those additional costs that have been imposed on the water services paid by the subscribers who really generate them. For instance, the extra investment needed to satisfy demand for water at the height of the tourist season. These costs should clearly be recovered through tourism. How many people realise that, in a temperate climate like that of France, an 18-hole golf course consumes as much water as a town of around 40,000 inhabitants. In an increasingly competitive tourist market, many communities have created such facilities in order to offer a wide range of activities to their summer visitors. The Mediterranean Sea is the world's most popular tourist destination. In the area around it, the population increases from 150 million in winter to 250 million in July and August. An average tourist there uses 300 litres of water a day, that is, twice as much as the permanent residents, and for luxury tourism this can rise to as much as 880 litres per person per day! In many coastal communities, differentiated tariff scales place a higher charge on hotels and other tourist facilities, making tourism bear a fairer proportion of the costs, a strategy that could usefully be copied elsewhere.

PERFORMANCE-BASED PAYMENT SYSTEMS

In this second model, the public authority pays the operator directly according to performance. This payment is made up of two elements: achieved objectives and quantities of water billed. According to the way it is rolled out in practice, this model could possibly free itself from the "user pays for water" principle: it would put in place a financing system for water services paid for by both taxpayer and user.

This formulation – concentrating more on quality and funded through mixed financing – already exists in a specific form in Indianapolis, a city with 1.1 million inhabitants, where Veolia Water manages the water and sanitation services. Our payment is made up of a fixed element and a variable one, which is dependent on performance. A number of indicators have been defined, against which performance can be measured. They reflect key areas such as water quality, environmental protection and management of capital and investments. For each indicator, a target performance level is fixed, together with a method for measuring target performances – for example: achieving 99.9 % compliance with the quality standards laid down by the Environmental Protection Agency, or getting a 90 % client satisfaction rate. This system of indicators and variable payment encourages the operator to improve performance. In this sort of system, which can also be applied to a public operator, if a water service wishes to increase its revenue, it is more important for it to achieve qualitative objectives than to sell a greater volume of water.

SEPARATING THE VOLUME SOLD FROM THE VOLUME PUMPED

New solutions need new financial models, just as new scarcities need new water resources. Recycled waste water is one of these. In waste water recycling, payment to the operator is proportional to the volume billed but, most importantly, it is separated from the volume initially drawn from nature. Consequently, the encouragement to sell more to the user that results from receipts based on the volume consumed no longer negates the objective of reducing the volume of fresh water pumped. It does not upset the financial stability of the service, nor the requirement to create a margin for investment, nor the conservation of resources.

New scarcities bring new production patterns, and a financial structure based on the destruction of limited resources is replaced by one that helps to transform unusable water into useful water. Economic models and human ingenuity are constructed on the notion of scarcity. When a commodity is abundant and freely available, the question of its price is not an issue. But once it becomes limited, then prices, supply and demand all arise, explains economist Bernard Maris[1]. Scarcity has always stimulated an extraordinary explosion of human activity. Mankind has organised itself to compensate for lack of commodities, giving birth to agriculture, livestock farming, bartering systems, etc. The same is true of water. The dynamic pairing of scarcity and innovation has resulted, in particular over the last two decades, in the rapid development of technology for waste water recycling and for sea water desalination.

INCREASING WATER PRODUCTIVITY FOR CONSUMERS

We do not always realise that it takes 400,000 litres of water to make a car, 11,000 litres for a pair of jeans and 1,300 litres for one of the tiny mobile phones that we all use. To produce 1 kilo of wheat, 1,500 litres of water are needed. Virtual water is a term used for the water involved in producing a commodity or a service. Industry and agriculture, which both consume huge amounts of water, have a crucial role to play in controlling this water use that their ultimate clients never see.

China produces twice as much rice per hectare as India with the same volume of water. For every unit of GDP produced, it is thought that China uses six times as much water as South Korea and ten times as much as Japan. Even if these figures are approximate, and if the structure of their economies goes some of the way to explaining these differences, they certainly signal an immense margin for improvement. People talk a great deal about "dematerialising" economic growth. We also need to talk about "dehydrating" it.

Living in a world where resources are scarce means that we must use each cubic metre of water as efficiently as possible. Waste water recycling is a fast track for increasing the productivity of water. Separating water use from water abstraction means that maximum use can be made of the same quantity. Reusing water before it is finally discharged back into the environment greatly increases the productivity of each cubic metre borrowed from nature. It applies the precepts of industrial ecology to a continuous and open cycle, that of water.

In passing, it might be useful to highlight one of the paradoxes of virtual water. We have already seen that, unlike oil, it is not economic to transport water over long distances. Consequently, a global market in water is not a solution that can be used to correct local imbalances. The overabundance of water in Canada cannot help the desert regions of

145

---

1. Cited by Bruno Ventelou, in *Au-delà de la rareté*, Paris, Albin Michel, 2001.

Chile. Nevertheless, buying finished products from regions rich in water can, indirectly, compensate for a lack of water in other regions. For example, when a country imports cereals, it also imports the water contained in these agricultural products. While wet countries export their water virtually to drier countries, the reverse, surprisingly, is also true: countries which are arid, and sometimes poor, export their water to rich, rainy countries. This is the case when Egypt sells oranges and grapefruit to European countries.

THE END OF THE PRINCIPLE OF "USER PAYS FOR WATER"?

New ways of generating income to finance new tasks, a new billing and receipts structure to harmonise more with the true costs structure, new water cycles that will increase its productivity: some of these changes are still at the discussion stage, others are already being put into practice. Waste water recycling is spreading rapidly across the United States, Australia, Spain, etc. The other economic models which have been sketched out, those which combine payment through water bills and through taxes, depend on agreement in public authorities, perhaps even legislative change. For them to take shape, both legislators and consumers must accept them.

We would like to make these new or little-used models more widespread, but is this what our partners want? Some say yes, others no. How can we dehydrate production when the producer can waste water with impunity because he is not paying for it, and the client does not pay the cost of the water because it is not reflected in the price of the product? How can we avoid adding to the price of drinking water costs that are not directly connected with it when public authorities oppose the idea, and further claim that financing projects through bills is less painful than financing them through taxes?

In the end, applying the "user pays for water" principle depends on what consumers are prepared to pay and what public authorities are prepared to ask them to pay. For many years, some sections of the international community have regarded this principle as sacred, compelling water to work as a closed circuit. The whole cost of the service had to be recouped in the price of water. This solution poses problems, as we have seen. In developed countries, on the one hand, the expression "user pays for water" does not say how far the boundary of the principle extends. Should we include all the costs of depollution, of the restoration of aquatic environments, of social imperatives? This is a topic for discussion. In less advanced countries, on the other hand, this solution is not socially acceptable: it is not realistic to place all the costs of constructing new infrastructures on the consumer. Access to water must then rest on different economic models.

In France, most water professionals recognise that new financial resources must be found to modernise infrastructures. There are still costly investments to be made, both for finishing the enormous work of upgrading the sanitation system begun in the 1990s, and for following the 2000 European Directive calling for the restoration of the quality of water in the environment. To this must be added the investments imposed by the likelihood of an increasing scarcity of water in the south of France.

This point of view leads to an acid test. As the recovery of costs is based on water consumption that is structurally destined to fall, we will need to re-examine the financial structures associated with water. There are several possible solutions in France. Some people think that the price of water cannot be raised because consumers, always very sensitive to price rises, would oppose it, especially in a context of weakened purchasing power. It is not a question of the price itself, which is objectively low, especially compared to other European countries, so much as the consumers' perception of it. Events are driven by the belief that the price has always been too high. On this view, consumer sensitivity would be too great for prices to be allowed to rise, and few politicians would risk forcing a rise in their constituencies. If the "user pays for water" principle is not applied – even to those elements that are strictly related to the water service (any others having to be financed through taxes) – then it will be necessary to obtain finance from outside the water services. While allowing the modernisation of infrastructure, this option would be invisible to consumers and would avoid protests from them, but it would also take away their responsibility and make the pricing of water less transparent.

Should we raise this question as to whether all costs should be covered by the price of water? Rather than abolish one of the founding principles of water economics in Europe, it is better to continue applying it (even if it means accepting increased tariffs), while at the same time ensuring that projects outside the remit of the water service are financed by the taxpayer. We must collectively accept the cost of water services. In exchange, water professionals, whether local communities, water authorities, related bodies, or public and private operators, should accept, in the most transparent way possible, that they must provide a better quality of service, that is to say, accept higher costs and thus a higher price, if the whole cost is to be reflected in the price.

147

A further option would be to refuse to make this choice, and to refuse either to raise the price of water or to establish complementary financing from outside the water service. This would result in postponing vital investment and improvements in standards. But yesterday's non-decisions have to be paid for today! The European Court has warned 11 countries for not conforming to the Directive on urban waste water treatment, passed 17 years ago in 1991. There are 121 French public authorities still in contravention, and France is threatened with a fine of €400 million. In the same way, what is popular with voters today brings with it the risk, in some communities, of brutal price rises tomorrow, when it becomes a matter of urgency to provide the compulsory investments and to make up for lost time. It also risks placing water and sanitation services in breech of law, and in danger of lowering their quality.

III. HELPING MORE THE DISADVANTAGED

Mechanisms for assistance need support, especially as they apply to people with no access to drinking water and sanitation services, living in emerging or developing countries. These countries are moving, at varying speeds, towards the Millennium Development Goals, and, beyond this medium-term objective, towards an acceptable service for all their people. In developed countries, greater economic insecurity compels governments to find ways to maintain access to water for poor people. Among the approaches that have been examined, some are new, some have been long tried and tested, but all need to be strengthened.

DEVELOPING COUNTRIES: WATER AND SANITATION FOR THE GREATEST NUMBER

• Making services accessible

Setting up pricing policies suited to the inhabitants' ability to pay
Access to essential services demands that tariffs should be socially acceptable. Both water and electricity must be paid for in order to prevent waste and to ensure that the service is permanent and does not fail due to lack of income. But that does not mean that everyone should pay the same price. A differential pricing policy for different levels of consumption allows the cost of basic services to be reduced for the poorest people. In all countries of the world, it is public authorities who set the tariffs policy determining the price of water and the number and level of different blocks of consumption. Many countries have adopted progressive tariffs, with a reduced price for the bracket corresponding to the lowest consumption (called social tier). Even though they are clearly constrained by the threshold effect and the issue of connections being shared between several families, these tariff systems at least help some of the poorest users. They are found in a range of countries, including South Africa, Argentina, Algeria, Burkina Faso, Bolivia, Brazil, Colombia, Costa Rica, Ivory Coast, Guinea, Indonesia, Mali, Mauritania, Nicaragua, Panama, Paraguay, Peru, Senegal, Tunisia, Uruguay, and Venezuela.
In Gabon, the water service sells at a loss (that is, at less than cost price) in the social tier, which consists of users who consume less than

15 m³ per month. Water tariffs in Niger are among the lowest in the region, with an average price of 261 CFA francs (about €0.40) per m³ in 2007. The high rate of payment of bills (96 % for domestic customers and 99 % for businesses) testifies to the affordability of water prices. Chile has established a system of direct aid. It is financed by the taxpayer and so does not result in cross-subsidisation between users according to their level of consumption. This aid is given only to people who have been identified as poor by the public authorities, and eligible users first need to register in order to benefit. In practice, they receive a bill, but in order to pay it they are given vouchers, which they collect from the town hall. This gives them a reduction of between 25 % and 85 % of the price of the water and applies to the first 15 m³ of each monthly bill. It is calculated in such a way that the cost of water is no more than 5 % of a family's budget. The state compensates for this reduction by directly refinancing the water services. The Chilean system is thought to be reaching 95 % of potential beneficiaries.

Establishing a socially acceptable price for new connections

It is not enough to reduce the price of water, if people do not have access to the public network. For poor people who wish to become connected, the main obstacle is the cost of a branch-pipe. The average cost of connection for poor families is equivalent to three months' salary in Manilla and up to six months' salary in urban areas of Kenya. What is the point of a socially acceptable price for the cubic meter of water, if connection costs are unaffordable and continue to exclude poor people from being connected? Drinking water consumption is very often subsidised by a socially assisted tariff, but connection to the water network is not. In other words, people who are already connected benefit from subsidies, while those who are not connected – often the very poorest – do not!

Different mechanisms can make the cost of connection to the water network, or paying for it, socially acceptable. Some public authorities install branch-pipes free to needy families, as in the district of Belgaum, in Karnataka in southern India, while others install them at a very low price. In Ougadougou, the capital of Burkina Faso, the price was cut to less that one sixth, to make it affordable to people who were not connected. It has fallen from 150,000 CFA francs (€150) to 20,000 CFA francs (€20) and is now below the average monthly salary (30,000 CFA francs or €30). The difference between the contribution charged to inhabitants and the real cost is borne by the Office nationale de l'eau et l'assainisse-ment (ONEA), then by The World Bank. In Ougadougou, the number of connections grew by two and a half times between 2001 and 2007, i.e. from 41,000 to 107,000. They serve 600,000 newly connected people!

In December 2005, the Director General of the ONEA announced publicly that the price of the branch-pipes was "a real Christmas present". The impact of this campaign was such that today this price serves as a national reference figure. In Gabon, the Société d'energie et d'eau is subsidising the installation of drinking water socially assisted connections, and reducing the price to around two thirds of the normal price.

21. Change in the price of connection to the public network in Ougadougou

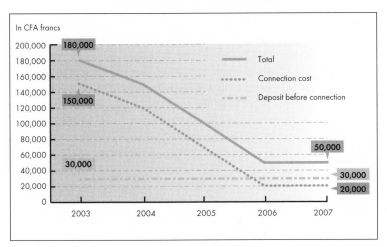

Other solutions concern the method of payment rather than the price of connection. In too many cities throughout the world, people wishing to be connected have to pay the whole cost of their connection in cash in a single payment, which is clearly impossible when the bill amounts to several months' salary! In order to remove this obstacle, REDAL (the company in charge of water, sanitation and electricity services in the Rabat-Salé conurbation in Morocco) has devised a novel way of paying for branch-pipes. It comprises a works fund, which finances extensions and work on socially assisted connections in advance; additional funds provided by public authorities if necessary. The payment for connection is then spread, without interest, over three to seven years. In total, in outlying deprived areas, more than 90 % of the inhabitants take up the offer of socially assisted connection. In Tangiers, too, a public works fund has been created specifically for socially assisted connections. It is funded from a charge collected by the communes, as well as from aid obtained from donors or by loans taken out with private banks. The main aim of this financial operation is to facilitate paying for connections while limiting the amount charged to beneficiaries of the scheme and spreading out the payments. The payment period varies from five to ten years, without interest, giving a maximum monthly payment of 100 dirhams (€9).

Enabling populations without formal property rights to be connected

Many public services will only deliver water to households with a legal right of ownership, yet more than a billion people live in urban and peri-urban areas that are not officially recognised! Abidjan, one of the richest cities of west Africa, has more than 80 unauthorised residential areas. Who is going to invest and build water infrastructures in such unstable peri-urban areas, where the inhabitants are considered illegal and can be evicted from one day to the next by the public authorities?

Nevertheless, change happens: it beats down the regulatory barriers that exclude public services. In Morocco, legislation banned operators from serving areas of unofficial housing. People excluded from public services and living in these areas were condemned to remain excluded. With the launch of the Initiative nationale pour le développement humain (INDH) in 2005, the government lifted this restriction. By authorising the supply of water, sanitation and electricity, the new public policy transformed the "illegal occupants" into "urban service users", with the rights and duties associated with this new status that official recognition had conferred on them. The INDH is prioritising the rebuilding of unregulated and under-equipped areas, by opening up roads, providing access to basic services and sorting out problems with land ownership. Public service operators are involved with this programme, which offers a structure to participants and marks a break with the previous urban policy. In Tangiers, Tetuan and Rabat-Salé, it was decided, after studies had been made of the areas where services were to be installed, that the socially assisted connections programme – originally planned to take place over 20 or more years – should be concentrated into five years. Three agreements fixed the objectives to be achieved by 2010 in these three conurbations in terms of access to essential services for districts and villages that did not have them already. For water and sanitation this affects 100,000 households, which is half a million people!

• Optimising and rationalising services

Optimising the functioning of infrastructures

On the door of the water services office in Majuro, a town of 25,000 inhabitants and capital of the Marshall Islands, there is a notice showing the days when water is available. Monday: on. Tuesday, Wednesday, Thursday: off. Friday: on. Saturday and Sunday: off. In total, clearly, the water service works for only two days a week. There are too many towns and cities in emerging countries that are, like Majuro, incapable of offering a continuous service to their population. The right to water, however, is hardly a passing luxury. Quite apart from any financial aspects, this situation comes back to a major issue: offering training in care and maintenance. It is no use paying for building infrastructure or buying state-of-the-art technology without providing the means to keep them in working order. In many towns and cities, including some in developed countries, drinking water networks are turned into sieves for want of maintenance. The provision of infrastructure in these places seems to follow an unchanging three-beat waltz: "construction – decay – reconstruction". In rural areas, the situation is hardly any better: in Burkina Faso, Mali, Malawi, Rwanda, Ethiopia and southern Asia, a third of rural water points are unusable through lack of maintenance, which cancels out the benefits which are supposed to have been brought by the infrastructures that states, communities, operators or NGOs have gone to the trouble of building. A "maintenance culture" has yet to be thought about or disseminated.

By operating the existing infrastructure to the utmost, it is possible to provide water to more people and still to reduce or defer some investments. This is, or should be, the main role of a professional water services operator. This can be done through such measures as adapting equipment instruction manuals, dividing the different stages of water distribution networks into sectors, and keeping up with day-to-day maintenance. Good management of services is an essential part of increasing the size of the network. The water thus saved can be redirected towards outlying districts, which often have the worst water supply. In Niger, drinking water supply has grown by nearly 8 % without investment in extending the capacity, simply thanks to a rationalisation of the system, which has allowed an extra 250,000 inhabitants to benefit.

Operators should put all their skill and knowledge into optimising infrastructure and operation, and, to do this, water services must adapt at all levels: technical, human, organisational, financial, environmental, etc. This "all-out" optimisation makes it possible to improve the quality of service provided, to generate recurring cash flow to finance new infrastructure, and, sometimes, to lower the price of water. When, in Gabon, average water and electricity bills fell by 17.25 % after a change to delegated management, it was thanks to rationalisation of purchasing and optimisation of the network, while at the same time freeing capital for investment at a much higher rate than had initially been envisaged. Since being contracted out in 1997, the Société d'énergie et d'eau du Gabon has invested more than 230 billion CFA francs, and another 400 billion of investment is planned ahead of schedule. This total of

more than 600 billion CFA francs is double the amount originally estimated in 1997.

Controlling costs

Low connection prices cannot be created without looking to control
costs. Reducing service costs is one of the most important factors in
connecting the maximum number of people with the same outlay, but
without compromising quality. In the long term, using low-quality materials will always turn out to be a false economy. In Ougadougou, the
introduction of new plumbing methods and new materials has led to
considerable savings in the cost of installing and maintaining water
connections. In addition, the meter repair workshop, created in 2003,
allows the re-use of existing materials, which brings down the cost of
meters by around 20 %. Large-scale operations, affecting entire districts,
also lead to economies of scale.

• Financing access to the service for the most deprived people

The speed of demographic growth is compelling many governments
in developing countries to build infrastructure on a very large scale.
Finance for water is currently insufficient, and it is in competition with
other equally vital needs such as energy, housing and transport. These
countries need to build both water and sanitation systems at the same
time, moreover, when in Europe they were constructed separately and
the work was spread over more than a century. It follows from this that
operators (whether public or private) have a pressing duty to manage
existing infrastructures in the best possible way and use any finance they
obtain in the most efficient way possible.

Combining several layers of financial support

The application of the "user pays for water" principle is unrealistic
in many developing cities and countries. If they are unable to recover the
whole cost of the service from users, public authorities need to organise
financial assistance, which can be organised in different ways: between
water service users, with a variable price structure in which the top consumption block could subsidise a socially assisted block; between large
centres of population and isolated areas; between taxpayers and users
(where water investment is paid for from the local authority budget);
between developed and emerging countries via international aid[1];

---

1. As is the case, for example, in the financing of the French Development Agency or the OBA
mechanisms of The World Bank.

between utilities, with electricity receipts, for instance, financing investment to be made in drinking water or sanitation.

Experience shows us the importance of combining different types of assistance together to make access to essential services as easy as possible. If we take the case of Gabon, three types of assistance have been superimposed to make the right to water into a reality: between users, which reduces the price of basic consumption; between large cities and isolated areas, with the former financing the latter; between utilities, with resources from the electricity service financing water service investment. All told, in the last ten years, the number of inhabitants connected to a modern drinking water supply service has grown from 40 % to nearly 70 % of the population.

When water services in developed countries cooperate with those in developing countries

Financial markets and institutions are creative entities, even if the financial crisis of 2008 has brutally shown us their limits. However, they do not appear to use much of their imagination on the question of water. A few happy but all too rare initiatives have emerged, such as the Oudin-Santini law relating to decentralised cooperation. This law, which was adopted in France on 9 February 2005, authorises local communities to allocate 1 % of their water and sanitation budget to financing emergency aid or cooperating on drinking water and sanitation projects abroad. The same applies to Water Agencies, which are financially autonomous public bodies. This law marked the renewal of decentralised cooperation over water. It sets up a financial circuit for water services in France to help towns and villages in developing countries to improve their drinking water supply and waste water collection and treatment systems. This mechanism has allowed the collection of €8 million in 2006 and €17 million in 2007, while an estimated €20 million have been committed for 2008. Six times this amount could be put to use each year if all the potential French participants (water services and public agencies) joined in.

This movement was started by the Syndicat des eaux d'Ile de France (SEDIF), which gathers 142 communes and supplies water to 4 million consumers. In 1986, it decided to finance development aid projects to improve access to drinking water in deprived areas. In this way, over 22 years, it has co-financed rural and urban water projects to the value of €13.5 million, which have benefited more than 2.2 million people in 17 countries, including Burkina Faso, Chad, Benin, Senegal, Madagascar, Vietnam, Cambodia, Laos, Haiti and the Comoros Islands. Today, SEDIF devotes €0.60 per m^3 of water to fieldwork projects carried out in developing countries with the support of NGOs. For an average Ile de France household, this reperesents less than one euro per year. In 2007, the total amount of aid gathered in this way reached €1.6 million, or 0.4 % of the water service's receipts.

Doubling finance for water

The extra investment needed to reach the Millennium Development Goals for water and sanitation have been estimated at $10 billion to $30 billion a year. It is possible to meet the challenge of financing the Millennium Goals if all the financial resources are increased, and not just some of them: public finance allocated to water in every country, contributions from users, local taxpayers' share, public development aid (of which only 5 % is allocated to water compared with 20 % to 30 % to telecommunications), multilateral agencies' contribution, cooperative projects between local communities, local savings banks, etc. Do we need to be reminded of this? But finance, especially when it comes as aid from abroad, cannot be increased if the conditions of governance that should guarantee that it will be well used are not met. And this is where we feel the pinch. Money is rarely the factor that prevents projects destined to improve water and sanitation services from emerging. The limiting factor depends much more on governance. Water crises, although they may appear to be environmental, are more often than not due to crises of governance. Without progress in governance, the hope that everyone can have access to water is merely a dream. Without good governance, there can be neither good technology nor good management of water services.

In this context, it is revealing to look at the finance currently in place to meet the Millennium Goals. On the one hand, one could say that the total sum involved is too large for the number of projects wich respect the criteria of good governance. On the other hand, this same sum would be far too small if there were enough good projects to achieve the Goals, projects that all fulfilled the criteria for good governance. This sum is even farther from being enough to achieve the real object, that of access for all, and not simply "half of all", to a genuine water and sanitation service. As Gérard Payen, president of AquaFed, said rather forcefully, "The Millennium Goals are only intermediate objectives, while populations expect much more." They are not only hoping for better wells, but also safe running water in the home. They do not only wish for private toilets, but also ask to be protected against potential contamination, which means waste water collection and disposal.

In developing countries, public finance will, for a long time, be the only real option for improving water and sanitation services, because its terms are very competitive, and can be extended for longer and are therefore more in tune with the economic lifespan of infrastructures that can be in use for half a century or more. Given the widespread nature of the needs, it is also essential to promote local finance on a large scale. This protects against fluctuations in the exchange rate, as the debt is then drawn up in the same currency as the income from the service. Since water and sanitation infrastructure are usually the responsibility of local communities, financial flows should be directed more towards them, rather than towards the state. Since these communities are "sub-sovereign" entities[2], financing them supposes that a "sub-sovereign" risk would be acceptable to lenders or investors. One of the tasks of central government is to encourage the expansion of local financial markets, so that they are mature enough to accept such a risk and thus provide long-term finance to the water sector.

There are those who would like to count on large private operators to finance the vast infrastructure programmes that are needed around the globe, but they fail to grasp two fundamental points. Firstly, on a global scale, the part played by private operators in water and sanitation services is marginal: they supply to less than 10 % of the world population. It is thus futile to hope that they will be able to finance the huge investments needed for water. Secondly, this is not their role; an operator is not a banker! Businesses, when they are performing well, do so because they carry out their work with real skill and knowledge, without straying from their field of expertise. The mission of an operator is to manage the infrastructure for which he is responsible, not to finance it. Even if private operators, as part of a contractual arrangement, are able to finance part of an investment within the limit of their available cash flows, their role is mainly that of a catalyst in raising funds from financial partners and making sure they are put to good use. When they bring a service up to standard, they are preparing the ground for good use to be made of the finance that has been raised.

• Weaving together new cooperative enterprises

Tailoring technological solutions to the needs of disadvantaged populations is essential, if their demands are to be met, but, despite its importance, technological innovation alone is generally not enough. It needs to be matched by organisational and financial innovation so as to create an entity that is capable of responding to all the needs of poor families in the long term.

These days, there are many initiatives proposing new ways of solving classic economic equations. It is doubtless too early to claim with any certainty that these new models are permanent and will provide an effi-

----
2. Organisations below state level, such as local public authorities.

cient way of improving access to water, but the creativity demonstrated by their promoters should be welcomed, and should be better explained.

An entrepreneurial model to provide basic sanitation to the poor:
Sulabh in India

In India, the Sulabh organisation has established basic sanitation services aimed at the poorest members of society, namely the lower castes and migrant workers. Founded in 1970, this association has, by its own efforts, now become one of the world's largest NGOs supplying basic sanitary equipment. By facilitating the construction of more than 7,500 public toilet blocks and 1.2 million toilets at a low cost (of between $10 and $500), it has brought basic sanitation to 10 million people. These services have a considerable impact on the lives of the inhabitants, even if the systems are not complete, and do not include the collection and treatment of waste water. Originating in the state of Bihar, Sulabh today operates in 27 of the Indian states as well as in Sri Lanka, Bhutan and South Africa. The success of the organisation, and the level it has now reached, can be attributed to the way that it adapts its services to poor populations, as well as to the commercial viability of its products. "Sulabh follows a business not a charity model. It enters into contracts with municipalities and public sector providers to construct toilet blocks with public funds. Local authorities provide land, and finance the initial connections to utility services, but all recurrent costs are financed through user charges. Fees are set at about 1 rupee (2 US cents)[3]." In return, Sulabh undertakes the long-term management of the equipment, which can be for as long as 30 years. Service costs are financed by the individual user charges, with free access for children, people with disabilities and those who cannot afford to pay. One of the merits of Sulabh is that it provides a wider service. "Sulabh complexes" installed in shanty-towns bring a much appreciated, multiple function service: latrines, showers and, occasionally, medical visits.

Sulabh illustrates one of the most fundamental characteristics of public water services: the importance of partnership. Essential services are a sector in which no participant can work alone, especially in developing countries, and most especially if all are to have access to clean drinking water. In areas of urban poverty, the implementation of water supply and sanitation policies is only possible with cooperation between public authorities, service operators, NGOs and financial providers. Other organisations can also play an intermediary role in bringing people together. Experience shows that far-reaching social policy cannot be carried out with limited partnerships. It is essential to create new types

WATER

FINDING NEW MODELS

---

3. Human Development Report 2006, UNDP, *op. cit.*

of collaboration by marrying together the skills and knowledge of all the parties.

"No loss, no dividend", a new doctrine for new businesses?

At first, nature seems to have been generous to Bangladesh. It is one of the richest countries in water. Its subsoil has abundant groundwater sources that are not too deep, and therefore provide relatively easy access to water. Unfortunately, for geological reasons, almost all of the groundwater has been found to be contaminated with arsenic, sometimes at levels that make it a health hazard. Today more than 30 million Bangladeshis are exposed to the sometimes fatal consequences of chronic arsenic poisoning.

It was in this context that a meeting took place between Grameen Bank and Veolia Water. Each of these organisations has complementary skills and knowledge about supplying drinking water to the most deprived rural communities in Bangladesh. For Grameen Bank, "the village bank", this is a chance to apply to drinking water the "social business" principles established by its founder, Muhammad Yunus, winner of the Nobel Peace Prize in 2006. Under his leadership, this bank, originally specialising in microfinance, has become a globally renowned institution. For Veolia Water it is a chance to test out solutions for the efficient supply of drinking water to disadvantaged rural communities, while at the same time meeting two challenges. The first is technological: dealing with the presence of arsenic in groundwater. The second is social, and arises from the limited financial resources of the populations concerned. It is, then, an opportunity to explore new economic and working models for contributing to the Millennium Goals.

The alliance of Muhammad Yunus's "social business" theory and Veolia Water's expertise has resulted in the creation of a joint enterprise. Financed 50 % by Veolia Water A.M.I. and 50 % by Grameen Healthcare, this company will bring clean drinking water to 100,000 inhabitants, for a total investment estimated at €500,000. Veolia Water is responsible for constructing the facilities for producing drinking water from surface water resources, for ensuring their maintenance and for training the workforce. Grameen Bank's role is to negotiate with villages, to draw up a management and price-structure model suitable for the population concerned, and to deal with water distribution. For Muhammad Yunus, "Economy must adapt itself to the needs of the poor, and to begin with, provide for their essential needs such as the need for drinking water. Grameen-Veolia Water Ltd. is dedicated to this aim, and I expect a lot from this partnership." Muhammad Yunus is a busy man; he is busy doing good. When he believes an idea to be feasible and beneficial to poor people, he wants to try it out on a grand scale as quickly as possible, so the work schedule of the joint-venture company, founded in March 2008, is tight. The delivery of the first production unit is planned for the beginning of 2009. It will supply drinking water to the inhabitants of Goalmari, a village 100 km from Dhaka, and will create around ten jobs. The investment will be gradually paid back through sales of water. Previously free, but poisoned, the water will be sold at around €0.15 for ten litres, a price set by Grameen Bank taking into account the villagers' financial means. No charity towards poor populations, but no dividends for company shareholders, either. Grameen-Veolia Water Ltd will work on the "social business" principle: "no loss, no dividend". Consequently, all the profits will be reinvested in the project in order to finance its expansion or to replicate it elsewhere. By bringing an immediate operational response to a vital need, this company will be helping to integrate poor populations into the real economy, instead of excluding them. What we might term "social business" does not abolish the market economy or the role of business, but adapts them in order to meet the basic needs of disadvantaged people. Far from flouting business viability, the company harnesses it to improve conditions for the poor.

This new type of economic partnership is diversifying the world of water. It associates development organisation with private enterprise in a hybrid model, so as to maximise the social benefits. It aims to bring a permanent solution, within viable economic conditions. This is why it does not depend on any subsidies, which, by their nature, tend to be short-lived. No system can be self-sustaining if it depends on aid that might be removed from one day to the next, a fact painfully borne out by the drop in public development aid reported in 2007 by OECD. Once more, the major donor countries have not honoured their commitments.

This project is just beginning. Will it bear fruit? How will it stand the test of time? Will it be applicable elsewhere? Only time can tell, and it is still too soon for the answers, but in 2010 we should learn the first lessons from it. Like any innovation that is both social and economic, this project will probably need adjustment over the first few years of its life.

DEVELOPED COUNTRIES: MAINTAINING ACCESS TO SERVICES FOR THE POOREST

In developed countries, the issue is not one of creating new connections, as it is in developing countries, but of maintaining access to water services for people who are already connected but are facing hardship. In theory, there are three major options: reducing consumption to the essential so as to help people manage their bills (using prepayment meters); adapting pricing structures so as to transfer the cost between different types of user; and applying various social aid measures financed by other users or by taxpayers. The first of these options is, in fact, little used.

• Less increasing block tariffs but more direct individual aid

In developed countries, socially-assisted tariffs with cross-subsidies between the higher blocks and the lower blocks, are less usual than in developing countries. One reason is that a smaller proportion of the population experiences hardship; another is that these countries prefer systems of direct aid, which target needy individuals better, even if they are more expensive to manage.

Some communities have nevertheless brought in a reduced tariff for part of what is consumed, thus helping to maintain a water service to people on low incomes. In France, the Rouen conurbation now charges a reduced tariff on the first 60 m^3 of water billed; this facility results in a saving of around 30 % on consumers' water bills. Progressive pricing can also be found in Spain, Greece, Portugal, Italy, the United States and Japan; it creates a financial interdependence between large-scale and small-scale consumers of water, in such a way as to reduce the price of the first few cubic metres on each bill. The reduction can also be applied to the standing charge.

Other more complex pricing structures take into account the size of the family. In the United Kingdom, Luxembourg and Malta, this type of arrangement is used. Other countries have established preferential pricing, but unlike progressive pricing, the tariffs only apply to the poorest users. This mechanism combines the cross-subsidies of progressive pricing with the precise targeting of recipients found in direct social aid mechanisms. Thus, in Flanders, a system has been instituted whereby all domestic customers benefit from a socially assisted tariff for up to 15 m^3 per year, but people in difficult circumstances do not pay it, and so receive that volume of water at no cost. Some countries, for example Australia, take state of health into account in order to reduce the water bill of people suffering from illnesses where the treatment requires large quantities of water.

In almost all countries, those in financial difficulty are first given an extension of the payment date of their bills. In France, the Fédération professionelle des entreprises de l'eau (FP2E), whose members supply 72 % of the population, records more than 500,000 unpaid bills every year. To anticipate this kind of situation, people who wish to may pay monthly, as this is a better payment method for those on low incomes.

In France, outstanding debts amount to 0.7 % of the total billed. In England and Wales, they reached 1.7 % in 2006-2007.

When customers in a difficult financial situation cannot pay their water bills at all, help takes the form of a partial or total cancelling of their debt. This is the case in Belgium (in Wallonia and in Brussels) and in the United Kingdom. It is also the case in France, thanks to the "water" component of the Housing Solidarity Fund instituted by the government. This constituent allows the bills of people on a low income to be taken over. In practice, the water service operator, the public authority, the Water Agency and the state all abandon their respective shares of the debt. This solidarity system now operates in 70 départements. In 2006, the Housing Solidarity Fund, helped nearly 50,000 customers with their water bills amounting to a total of €6.3 million. This strategy, which started in 1996, was originally optional, but has lately been given legal force by the French government. This has led to two major consequences: firstly, it is now obligatory for all départements, and secondly, the "one-stop shop" has emerged to deal with help given in paying for all utilities (telephone, housing, gas, electricity and water).

• Individual billing in housing blocks, an idea with difficulties

Some public authorities encourage people living in housing blocks to have individual meters fitted to be able to put in place a socially assisted pricing policy that will benefit the greatest number. This can turn out not to be a such a good idea as it seemed, as is shown by the study carried out for the City of Paris in 2007 by Bernard Barraqué, Research Director at the CNRS. The study looks at the installation of individual meters and the move towards individual billing. Just as they do for electricity or gas, each household would receive its own bill for water and pay for the amount consumed in their own apartment. According to its supporters, individual metering would reduce water consumption by making users more responsible and would lower the bills of disadvantaged groups. Having listened to the arguments of financial providers, and of some consumers, the City of Paris wanted to investigate more deeply.

"In Paris, water billing is calculated on the total consumption of the whole building, which is then divided between apartments according to their floor area," explains Bernard Barraqué. A review of the literature at first led them to "rule out the simple and widespread idea that fairness to consumers, i.e. each user paying for the volume of water consumed, would be compatible with social justice, i.e. payment for an essential public service like water, in a way that is proportional to, or at least compatible with, households' ability to pay."

Then the study sifted through actual cases where a change to individual metering had occurred in order to measure the effects. In Toulon, the change resulted in a rise of 30 % in bills. In addition, "the non-payment rate rose to 11 % in one of the blocks studied, considerably more than the national average"! In Paris an apartment block on the rue Notre-Dame de Nazareth was studied, but: "no significant drop in consumption can be observed, while the increase in the total of all the bills compared with the previous collective bill is around 30 %. This means that here, too, the gains of the winners are cancelled out by the cost of the new standing charges." In these blocks, the public water service, even if it becomes more expensive, is not valuable enough to merit such a detailed – and thus costly – procedure as individual metering.

• PIMMS: the new "one-stop shop" link to public services

In parallel with these financial systems, some new ideas are being tried out in towns to maintain the link between public services and the people. One interesting example of these is the PIMMS, which are the product of an association between public bodies and businesses (such as La Poste, EDF, France Telecom, Veolia Water, etc.). These centres aim to provide a focal point in sensitive urban areas, and to contribute towards the development of the local economy. In practice, they offer a variety of services: they provide general information about public services, including water; give advice on choosing between different types of service and using them effectively; help people find solutions to problems over payment of bills; introduce people to IT skills and help them use the internet or write a *cv*; and provide a local outlet for stamps, public transport tickets, etc.

The first PIMMS was opened in 1995, in Lyons, and since then they have spread all over France. Today there are 29, seven of which are in the Lyons conurbation. They employ 127 people and are very active, with 240,000 visitors recorded in 2007.

IV. GOVERNANCE: AT THE ROOT OF THE PROBLEM, AT THE HEART OF THE SOLUTION

There are various ways to cope with water problems better. They include alternative resources and innovative technologies, suppleness and adaptability in economic and financial systems, a new social eco-system and new forms of cooperation. These ways are already known and followed by many but it remains to follow them through, to extend these initiatives, which, however effective they may be, still remain too limited to deal with the scope of the challenges to be met. Although every situation is a local one, there is one common factor that explains the difficulties everywhere: weaknesses in governance.

The concept of governance has now become common coinage: it is invoked to solve all our evils. Not a day passes when it is not called upon in one sector or another, but this has not always been the case for a concept that is both so old and so new. Originally, governance simply meant a method or system of government. The word governance comes from "to govern", and is synonymous with management, administration, leadership. Governance has existed ever since there have been governments. In the beginning, the expression was applied to public affairs but then it was extended to the private sphere with the governance of businesses. However, with official recognition of sustainable development in the 1990s, its meaning has shifted: governance has become a way of managing public or private affairs that allows for the right of civil society to have scrutiny. The 1992 Rio de Janeiro Earth Summit gave its blessing to this development: its final declaration states not less than 27 principles, including "participation of all parties involved and of new governance." We could say that governance constitutes the totality of cross-mechanisms set up to optimise decisions for the sake of the common interest.

This concept is particularly pertinent to the water industry. It refers to three complementary, intimately linked ideas. Firstly, governance clarifies the interactions between the different players involved with water. This is a vital issue, since the proper distribution of roles in water management is essential for the success of the service. Secondly, governance refers to the dialogue that needs to take place between all the players; a consultation that is crucial, for "only what has been negotiated is sustainable in water matters[1]." Lastly, governance refers to the

---

1. Michel Camdessus, Kyoto World Water Forum, March 2003.

need for transparency in performance, and with delegated management, in contracts. The capacity to bring together these three elements (assignment of roles, dialogue, transparency) and to implement them, is what makes the notion of governance fundamental to water management.

THE PATHOLOGIES OF GOVERNANCE

The many water crises in the world bear witness, directly or indirectly, to failure of governance. As in medicine, where diseases generate information about good health and the conditions for maintaining it, water crises illustrate the conditions for good governance. Let us look at some of the main pathologies of governance as it relates to the water industry.

– Lack of distance between regulators and operators. Often it is up to local authorities to regulate the operator, whether public or private. But when the two opposing functions, the daily running of the service and the control of that operation, are not clearly distinguished or are carried out by the same body, the water service and its regulating authority become both judge and defendant. The control function inevitably suffers from this confusion and is weakened. It risks becoming complacent, to the detriment of both service users and the environment. Isn't that the case in some towns in the USA, where leakage in the distribution networks can reach 40 % to 50 % of the volume carried? Or in Europe, where some authorities have been slow to apply the 1991 Directive on urban waste water treatment, and make the countries to which they belong liable to a fine?

– "Under-regulation". Weak regulation of water services leads to a weak performance. The high number of public operators in developing countries who receive neither clear objectives nor the necessary means to serve their territory is a clear sign of the absence or ineffectiveness of regulatory mechanisms. When so many outer-urban districts are not served or are only supplied with water for a few hours per day, it becomes clear that the principle of a universal water service is being flouted. There is a real lack of regulation, at local, regional or national level.

– Agricultural water withdrawals and waste-dumping. Too often water governance stops at the edge of towns. Its agricultural dimension is neglected, with consequent over-exploitation of groundwater in many parts of the world. In Brittany, the pollution of resources by pesticides and nitrates is evidence of a failure of governance.

– Effluents from industry and small business. When, despite the standards laid down, great industrial factories, such as metallurgical or food processing plants and tanneries, dump their untreated waste water into Asian rivers, isn't that clear evidence of a lack of policing of industrial waste water?

– Management of trans-border rivers of aquifers. The fact that a river or underground water source runs through several countries adds substantially to the difficulty of creating an efficient system of governance. Indeed, there are no legal means to put pressure on a state that behaves like an environmental maverick. Thus the Aral Sea was sacrificed to the god of cotton, and continues to be so. The Baltic Sea, polluted by untreated effluents from the Baltic countries and Russia, regularly suffers proliferations of algae, to the detriment of the Scandinavian countries.

– "Over-regulation". We find this less often than other failures of governance. It takes the form of heavy administrative burdens imposed on operators, by a plethora of controlling bodies, who multiply demands for information, insist on permissions being obtained to perform tasks (which, in fact, relate to the everyday running of the service) and are guilty of excessive delays in granting authorisations. Too much energy is then spent on bureaucracy. The operator and residents have to bear the additional costs of these demands, which are disproportionate to what is delivered.

GOOD GOVERNANCE MEANS A PROPER ALLOCATION OF ROLES

• Subsidiarity: local government for a local service

Water is a product whose price is less than a euro cent per kg, but whose transport cost, because of its weight (a cubic metre of water weighs a metric ton) is very high. That is the reason why, with rare exceptions, it is sourced, distributed and treated near its point of use. Of course, different regions have very different conditions, because of their geographical position, their available water resources, and the nature and intensity of the human activities carried on in them. That explains the differences in investment levels and water price. That is why, by definition, water is a local public service. The local governments of the world strongly emphasised this at the 4th World Water Forum in Mexico in 2006: "Local governments play a fundamental role in the management of the water resource and in the organisation of public water and sanitation services. Their role must be recognised and strengthened. Local authorities must be able to choose freely between different management methods." "Local", then, must be at the heart of the governance system, which should be organised to suit the terrain, and the needs of municipalities and consumers.

A very large consensus emerged on this point after the World Water Forum in Mexico, whose central theme was to take local action for a global challenge. Adopting the fundamental principles of French water policy, in the year 2000 the European Water Framework Directive had already reaffirmed the need to work as closely as possible to the local set-up, according to the principle of subsidiarity. However, the local authority, if it exists, must have the human and financial means to shoulder its responsibilities. Michel Camdessus' message to financial institutions in 2003 was important: "Donors must be prepared to direct aid towards local authorities, who need loans at preferential rates for projects in the water sector." This message can only be translated into action if the right conditions for running the service have been established in advance between the different players.

• Rights and duties of public authorities and operators

Confronted with the same task, countries have chosen different organisations and forms of governance, but three constants emerge from this organisational diversity. Firstly, it has to be clear who is at the heart of the governance system. Is it the local authority as in Germany, France, Czech Republic, China and Australia? Is it a specialised agency such as OFWAT in England? Is it the state, or its intermediaries, as in Burkina Faso or Madagascar? In every country, a public authority is at the heart of governance, and in most countries, this public authority is local, because that is where the responsibility for providing a water and sanitation service falls. Secondly, good governance is not tied to a single form of organisation. It can work with direct management of public services, but also with indirect management, through a private operator, or with mixed management, as with mixed economy companies in France or joint venture businesses in China. It can also work with an operator dealing with very different areas, which may be local, national or regional. Thirdly, strong environmental policies cannot work with weak mechanisms of governance, and there is no truly efficacious governance without performance measurement. The efficacy of environmental policies depends strictly on their application and control.

It would be an illusion, sometimes a dangerous one, to think that individuals, ward associations, or local entrepreneurs could, on their own, fully replace a deficient State or local authority. Of course, the many operators in the informal economy are indispensable and will remain so for a long while, even if their activity often results in additional costs and poor quality water when compared to that delivered through the public network. In Latin America, Asia and Africa, water-sellers, water carriers, hydrant operators play a vital role for millions of people, despite the poor quality of the service they deliver. They fill the gaps left by the municipalities. For better or worse, they stop the gaps and make up for the deficiency of the public service. Their importance is in inverse proportion to the performance of official operators.

But allowing groups of residents to take their fate into their own hands in the matter of water distribution is certainly not satisfactory in towns, and even less so in the great cities. Here social links have expanded and the communitarian approach may end up with what the UN Habitat Programme observed in the slums of Kenya: worse access to drinking water for the poorest because of the rising costs of payments to "intermediaries". Primary governance is that of the state, whose role it is to define the general water policy framework and the conditions for subsidiarity. In Africa, the countries most in step with the Millennium Goals (Morocco, Uganda, South Africa, Gabon) are those who have defined a national water policy, and made it a priority. French water policy itself would not have been as successful as it is, if the 1964 law had not given it a frame of reference by the creation of "watershed financial agencies". That facilitated resource management and recognised the community as the appropriate level for the organisation of services. Once the national framework has been set up, local authorities can make their own local water and sanitation policy, and take on their role as the organising authority.

No water policy can work without a clear division of responsibilities between organising authorities, service operators (whatever their status), and funders. Good governance requires the formalisation of rights and duties between players, and a clear distinction between regulator and "regulated". Mixed roles, the fact that some do the job of others or that these jobs are not clearly defined, are a source of confusion and failure. When each does his own job well, a positive dynamism can emerge to improve essential services.

One of the major functions of the organising authority is constantly to urge the (public or private) operator to make progress, and to prevent his becoming content with mediocre performance. In the long term, there can be no efficient operator without a strong organising authority. It is up to local authorities to give themselves the means to control service performance (particularly the relation between price and quality), so as to be able to justify it to the user. That is why it is essential to create a general culture of performance evaluation. To take the case of France, performance indicators are defined by the decree and order of 2 May 2007, and they have to be included in the mayoral report. The most interesting point here is the analysis of trends from one year to the next. Creating a frame of reference for indicators, analysing the results over the long term, is part of the preparation for joint management.

Within public-private partnerships, the sharing of roles between the public authority and the water service manager is organised with great rigour, on the basis of a contract. This is one of the intrinsic virtues in these arrangements because it enforces a healthy separation of roles between the policy decision-making authority, on the one hand, and the operator, on the other. These partnerships encourage precise definitions of the results expected, and a focus on performance. In that way, they are an efficient means of establishing good governance.

NOTHING SUSTAINABLE CAN BE CREATED WITHOUT DIALOGUE AND CONSULTATION

Over the course of time, consumers have become demanding and legitimate players in the management of the public water service. Other players, organisations within civil society, have lately invited themselves into the debate. It is right that they should have a place in it. At a first stage, the initial dialogue between authority and operator mutated into a three-way conversation, demanded by consumers and their representatives. Then this triangular relationship "local authority – operator – customer" became transformed into a multiple relationship, with the inclusion of other parts of civil society, in particular, environmental protection associations. This has represented a major change in the governance of the water services, which has been strengthened by these new bonds.

Today, this multi-voiced debate is a matter for all. There is room for everyone concerned about water, since in the past there has been too much indifference. It is not just a policy that ensures success in the long term, but the viability of a debate in which concessions made by some ensure the progress of all. The historic dialogue between the public authority and the manager of a public service naturally remains at the heart of the consultation, but often it is not enough. It can even become a source of misunderstandings if the conditions for the operator's work lack transparency, especially if he is private and foreign.

Indeed, without consultations with the local population and consumers, nothing legitimate or sustainable can be built. It is vitally important that the values and policies decided upon by the public authorities and underlying essential services, sooner or later gain residents' consent. Citizen participation occurs, first of all, through the intermediary of their elected representatives and the public authorities representing them.

Everywhere in the world, the public authority – whether it be a nation state as in Niger, a regional state like Selangor in Malaysia, or a local authority like The Hague – has the essential power to define water policy and organise the participation of citizens in the way it wants. Where this participation is real and sincerely accepted, it helps the acceptance and smooth running of the water management policy.

- • The "discovery" of the consumer
  by public water services

Where this has not already been done, it is important to set consumers at the heart of the water service's operations. Indeed, consumers' agreement with the water policy, and the levels of service and price, constitutes a key factor in its success. Consumer relations have been profoundly transformed over the last 15 years and these changes are due to several factors. Firstly, consumers are showing ever-greater interest in the subject of water. For them, water has become, or perhaps become again, a challenge and they no longer hesitate – in particular through their associations – to demand more accountability from operators, both public and private. These demands are healthy and legitimate, and operators are forced to adapt to their customers' opinions. By the expectations they express and the pressure they exert on mayors and local authorities, "user-electors" also contribute to the good governance of the water services.

In France, the 2002 "local democracy" law makes it compulsory to set up consultative committees of public service users in communities with more than 10,000 inhabitants and in inter-community cooperation bodies with more than 50,000 inhabitants. These committees examine the relationship between the mayor and the service management, and give their opinion beforehand when delegation of the service is mooted. They are indispensable for building or extending dialogue, and also for ensuring local regulation. Many mayors have not yet set up these committees, which not only play a part in local democracy, but are a bulwark against demagogy, entrenched positions and mistrust between players, who often do not know each other. Recurrent media debates about the appropriate price for water would be tempered, if the question had been analysed locally within these committees. It is at local level that any claims of "overbilling" for water could be checked, as could a situation of under charging, which is more often the case than realised, and ultimately leads to the tacit transferral to future generations of investment spending that is really necessary today. In the United States, the public regulator organises public meetings whenever necessary, and does so systematically when there is a demand for price changing. These meetings are open to all and everyone can express their opinion. Unfortunately, these meetings can sometimes become derailed, by being packed with opponents of the sitting majority, who use the space as a platform for an all-out offensive. Such shenanigans do not help the real dialogue water management needs.

- ### • Civil society's growing role in the regulation of public services

For some years now, the number of players taking part in the water debate has been growing. The sometimes hermetic world of "water professionals" and politicians has seen them arrive, often with some amazement. The emergence of these "non-contractual involved parties", as it has been agreed to call them, is good news, even though it has taken time to get to know one another and learn to converse together. These new participants enforce a complementary critical examination of the running of the service; they oblige everyone to explain more clearly what they do, and to build new relationships. This is an important element in the water management of today and tomorrow: public authorities, businesses, trade unions, consumers, NGOs must work ever more closely together. A water policy that wants to avoid misunderstandings and be accepted, and therefore sustainable, must be negotiated between a minimum of four key players in governance: public authorities, operators, consumers and civil society.

An experience that took place in Niger illustrates the virtue of these consultations. In October 2007, about 40 people who were active in civil society were invited by Veolia Water to take part in a day of discussions. The occasion was a mid-term assessment of the contract between the State of Niger and the company running the Niger water service. The aim of the day was to create a framework for dialogue, allowing representatives of consumers and local NGOs to understand the roles and responsibilities of each party better. The discussions were about the governance of water, the environment, community participation, water quality, and so on. Site visits to the drinking water production plant and to water hydrants were organised. The participants commended this initiative and declared their willingness to continue the dialogue and cooperation. Such consultations are indispensable. In the first place, and to use French terminology, citizens are not simply "administrees" – those to whom the service is administered – but users with legitimate expectations of information and understanding. Moreover, and perhaps above all, sharing in water management means becoming involved in a social bond. In that sense, the multiplication of round tables and information days are exercises in democracy.

Dialogue and consultation with consumers and civil society are the conditions for a service that is effective (because it is founded on the needs expressed by users) and legitimate (because it is accepted by the population). Without investigating consumer satisfaction, it is impossible really to know the aspirations they may express, and the progress they expect from the water service. These sometimes differ from the aims fixed by the public authorities, if not in their final outcomes, at least in their priorities. If opinion surveys of water service customers are ignored, it is easy to be deluded about their degree of satisfaction. In poorer wards of developing countries, mediation is indispensable. Without the involvement of ward associations as a means of consultation, without a team dedicated to it, the social connections programme in Morocco, for example, would never have been achieved to such a large extent, and the mobile agencies for inhabitants in districts remote from town centres would not have been set up.

Currently, we are regularly reminded that there is still some way to go to reach harmonious relationships between all the parties involved. It is very difficult to deal with such a complex issue as water in a context that fosters emotion, passion and manicheeism. But collaborating with new partners means building a positive debate around objective, shared information. It also requires a collective will to forge a responsible relationship, which rejects demagogy and caricature, the "infantile" evils of water governance.

• Shifting positions for water service management

Modern water management involves the creation of new social bonds, openness to hitherto unexpected partners, beyond the classic contractual relationships. Working with water forces people to rediscover the interest of extended collective games, and sometimes it shows its capacity to "make society". This need to create "new social bonds", particularly in emerging countries, means changing the way water services are run. We should not underestimate the extent of the internal changes that this new social context invites us to make. Both private and public operators still have a lot to learn on the subject.

Firstly, it has to do with time. Dialogue, consultation, partnerships with local associations and NGOs take time. Time is needed to set them up, then to make them work and measure the results. A long time-span is a normal operating framework for water service operators, but they can still find it quite difficult to agree to devote all the time necessary for dialogue and consultation. It is a key element that needs to be better respected.

Secondly, it concerns our relation to knowledge. Operators, particularly private operators, have long based their legitimacy on their expertise and technological mastery. That is perfectly understandable. Their knowledge is the main reason justifying a water companies' operating licence, and it is one of their trump cards in their competition with each other. In developing countries, where we are addressing populations without access to water, we sometimes need to "unlearn", because the knowledge of needs and solutions partly lies with the consumers themselves. Of course, the demand for water and sanitation is universal, but the feasibility of projects, especially from the social point of view, is often understood just as well – or even better – by the local populations than by our engineers. These populations must become players in the service. A real change of perspective is needed here. It is no longer a question of protecting one's knowledge in order to increase one's value in the economic competition. One needs to rely on available knowledge in the public space, which lies with the local population and its representatives, in order to work out together the most appropriate solutions for the area, thus combining technical know-how and local experience.

Lastly, it concerns our relation to human resources. Because the water management model in emerging countries, particularly for the poorest populations, is specific, we must train or recruit people capable of carrying out suitable consultation programmes. Building partnerships with civil society is a special job, and so it must be done professionally. That means we must think about who are the best qualified people, and the best way of attracting them. More broadly, all our staff need to share social concerns. This can be a considerable challenge, and means that we have to make them more aware of those concerns, understand them, and take them into account.

TRANSPARENCY: BUILDING A LONG-TERM RELATIONSHIP OF TRUST

"Anyone who wants to be respected must be transparent. Once, calling someone transparent was not a compliment. Now, accusing someone of not being transparent is insulting[2]," Guy Carcassonne, professor of constitutional law wrote in 2001.

----

2. Guy Carcassonne, "Le trouble de la transparence", *Pouvoirs* journal, no. 97, 2001.

• Encouraging transparency and performance assessment

The question of price differences rightly interests water service customers, but they are equally concerned about transparency and performance assessment. It is then up to each player in the water industry to satisfy the increasing demands for information.

In France, information about water quality has progressed a lot over the past few years. More and more precise data have been included in mayors' annual reports on the price and quality of the water and sanitation service. They are based on an array of performance indicators, not just for private operators but also for public ones. In its 2006 study comparing French water services under public management and those under delegated management, Boston Consulting Group noted that the number of performance indicators given by delegated managements is 24, as against an average of 14 for services under direct public management. Access to environmental information has become a constitutional right, set out in article 7 of the Environment Charter promulgated on 1 March 2005. The willingness to be accountable is also expressed in service commitments to customers, the creation of customer service centres and the carrying out of satisfaction surveys among consumers.

Even though these improvements are both real and perceived, the effort towards transparency needs to go further. French people want more information; henceforth the quality of information about water matters to them as much as the water quality itself. This information must be based on objective performance assessments of the water service, and then shared and used to inform the debate. That is the famous triptych: "evaluation – information – dialogue", which we mentioned earlier. The trustworthiness and objectivity of information must be based on rigorous assessment of water service performance, and the existence of an autonomous information authority, independent of the operators, whether public or private. Failing this, the information supplied may be doubted and accused of being partisan, as, for example, if it fails to give any account of the real service performance and only deals with prices. That would be to ignore the conditions for running a service, such as the quality of raw water resources or the topography of the distribution network, as though these were identical from one town to another. It would also assume that all services met sanitation standards to the same degree and with the same reliability.

Measuring performance is both necessary for transparent management of the water service and an assurance of progress. For private operators, it is also a contractual obligation. In this context, assessments carried out by external assessors play a key part. As well as legally required assessments, they add extra ones, enabling people to know the service results and to make progress. They also make possible a better quality dialogue between the parties, and provide a better knowledge of their aspirations. Taking account of these aspirations is vital for the long-term success of water services. That was the spirit of the social responsibility audit conducted in 2007 by VIGEO at the request of Veolia Water AMI (Africa Middle East India). It aimed to assess five main areas of social responsibility: attitude to markets, human resources, environmental protection, human rights protection, and social engagement. The audit was conducted by on-site investigations and conversations with different parties involved, both internal and external. It provided a great deal of information for the improvement of water services.

• Promoting ethics

For Gaston Bachelard, water is a symbol of purity, profaning it incurs people's rage: the polluter is a profaner. Every water or sanitation service operator must face the ethical responsibility arising out of what water represents in both the real and the imaginative lives of inhabitants, in all its many dimensions (mythical, symbolic, philosophical, spiritual and cultural).

First and foremost, ethics is eminently personal. Doubtless, there are collective ethics, but they cannot grow where there are no individual ethics. Ethics requires us to do what our conscience tells us. It means serving others and not just "helping oneself". Ethical behaviour is not satisfied by *a priori* or forced indignation, but seeks to be informed. It will not disregard the facts or ignore the reality on the ground. However, it will accept being questioned, which in the water industry, as in other sectors, is less frequent than may be expected.

Doing what one thinks one ought to do means regularly checking that one's actions are well founded. That requires humility and sometimes the courage to refuse to implement a bad policy. An ethical code cannot exist without sincerity and intellectual honesty. It means assuming all responsibilities and it governs action. It engenders trust. It is inseparable from the notion of duty: duty to oneself, to others, to those to whom commitments have been made, to those who will come after us, as well as those who have gone before us, from whom we have received so much; also a duty towards the environment, which enables us all to live.

Whatever organisations one belongs to, from the smallest humanitarian association to the largest institution, from the smallest public body to the most powerful state, from the humblest family business to the greatest private multinational, no system of governance, no ethical charter, no audit committee, can be a substitute for personal conscience and individual uprightness.

Ethics is also involved when social cohesion is needed. Nearly a billion people in the world have no access to drinking water, and more than two billion have no basic sanitation. That is the situation announced and denounced by the United Nations. Hence the first ethical claim is on the international community. We must respect the Millennium Goals for humanity entered into for 2015 and we should not lose sight of the longer-term objective, constantly reaffirmed, of access to water and sanitation for all, and not just half of those who are without today, as the Millennium Goals stipulate.

A desire for social and economic efficiency, for the connection to networks of those who are excluded, for cost recovery, means putting an end to practices like bribing meter-readers, altering meters to lower water bills, creating "special privileges" to discharge industrial waste water without treating it, and corruption in public markets, etc. A recent initiative should be commended particularly: the creation of the "water integrity network", whose secretariat is backed by Transparency International, one of the most active associations in the fight against corruption. Set up in 2006, this network aims to promote greater transparency in contracts for water services, both in developed and developing countries. AquaFed, which groups together all sizes of private operators in about 40 countries, is a member. If all the players involved with water, whether public, private, or involved through civil society, cooperate, then these challenges can be met. Here, as elsewhere, isolated action cannot bring about a global solution.

The ethical question is at the heart of the legitimacy of all water services. It is addressed to operators, but together with them, it is also addressed to politicians, because it is they who decide priorities, set the goals to be reached, and define the water price. They also choose the water management method (and incorporate criteria for the best ethical attitude, as was the case when responsibility for standpipes was decided in Niger). Together with service quality, ethical behaviour is one of the conditions for legitimacy and therefore trust. Finally, trust is one of the key words. No water service can be efficiently managed on the long run without trust. No partnership can be set up without trust. No finance can be raised without trust, and one of the principal aims of a system of governance is to create conditions for mutual trust. That involves a clear division of roles between local authorities, manufacturers, funders, operators, and so on. It requires transparency in procedures for transmitting information, a strict respect for roles, permanent ethical concern, and clear performance assessment criteria. Trust cannot be decreed. It has to be earned.

CONCLUSION

TOWARDS A CULTURE OF RESPONSIBILITY

In these last ten years, people have become aware of how water supply relates to issues like health, environment, development and peace. People who work in the water sector have turned their attention to integrated resource management, the spread of water scarcity, the financing of access to water and sanitation for all, the right to water, and the local management of services.

The 5th World Water Forum, which will take place in Istanbul in 2009, invites water specialists and policy makers to come together so that they can find a way to respond to global changes affecting the water sector: the population explosion, urban growth, changing lifestyles, climate change, the food crisis and land use. This meeting should remind us of what is glaringly obvious: what we most lack is not water resources, but the will to do something. More often than not, the resource exists, but what is missing is pipes! If we truly want it, access to water and sanitation for all will soon be a reality. Many examples can be cited showing that when political will and good governance have joined forces, the money required to direct water towards those who need it has always been found.

The time has come to think again about theory and practice. Ten years ago, in Europe, we saw the beginnings of a new culture, a way of thinking that shows respect to a resource that has become scarcer and more fragile. Going beyond the invitation to "waste less and pollute less", the emerging idea that water should be managed in a new, highly rational way by all who use it shaped the emergence of a new vision for water as part of an approach to sustainable development.

This new culture does not mean the end of the culture of engineers – I am one, myself – who for so long have used their knowledge to increase supply in order to meet the growing needs of human activity. Nor can it turn its back on the sanitary movement that has worked so hard for the public health of our citizens. Historically, public health has been the central issue for our profession, to the point of having inspired its economic model, and so it remains in emerging and developing countries.

A real change in culture will not be achieved by setting the old concerns against the new, but by integrating the new problem areas so as to reconstruct the culture of the water industry, and to accept that there is no single answer. The truly rational approach consists in recognising the appropriateness of solutions adapted to a region and compatible with its resources, local expectations and the level of development. This is a positive debate around shared knowledge and demonstrates a collective will to establish a mature way of thinking that will take us into the future.

The subject of water evokes so many strong feelings that this call for rational thinking may be frustrating to some people, but those who favour a romantic approach to water, promoting fantasy, ideology, confusion and untruths, are not serving the cause well. Lucidity and action are what is needed. By appealing to people's reason, intelligence and ingenuity, as well as to their good will, which is indispensable, water can remain what it has always been in the history of humanity: a link between people, and a link between nature and mankind.

The new culture of water will be one of responsibility. Not that there has been none in the past, but the breadth of the issues in the new century underlines and extends the responsibilities of everyone who plays a part.

The responsibility of the international community, which cannot escape from the commitments it made to humanity as part of the Millennium Goals. The responsibility of those who govern to make water and sanitation top national priorities in the many regions where everyone still does not have access to these two essential services. The responsibility of public authorities who can neither exempt themselves from decisions and schedules which they have undertaken, nor hide the cost of these projects, nor free themselves from the obligation to establish systems of governance which are practical and lead to action.

The responsibility of water and sanitation service operators, whether public or private. We operators have many responsibilities. We have at least a triple duty of efficiency, transparency and integrity. Society's demands have extended our professional boundaries. Not only are we responsible to our clients, we are also responsible to citizens, consumers and civil society. This book is a measure of that reality.

The responsibility of the financial community to accept municipal risks over a long term so as to finance construction of water infrastructure, as soon as conditions for good local governance have been fulfilled. The responsibility of domestic, industrial and agricultural consumers not to waste a scarce resource and to preserve it for this generation as well as for the future generations. The responsibility of industry and urban areas to protect the environment by cleaning the waste water they both discharge. The responsibility of civil society to be the true vehicle of social demand, to be the thorn in the side of public authorities and to act in the field, forgetting ideology, in order to improve water and sanitation services. The responsibility of everyone towards those who do not have access to essential services, by shouldering the cost of the right to water on behalf of those who cannot pay the whole cost themselves.

Since the time of Antoine Lavoisier, the father of modern chemistry, we have known that the only unremarkable thing about water is its appearance. Water management is an area where everything depends on everything else: as a system, its efficiency depends not only on the efficiency of each participant, but also on the links binding them together. "Everyone is responsible to everyone for everything," said Dostoyevsky[1]. Only shared responsibility for this common good that is water will allow us to reach its new frontiers.

----
1. Dostoyevsky, *The Brothers Karamazov*, 1880.